Wavelet Transform and Complexity

Edited by Dumitru Baleanu

Published in London, United Kingdom

IntechOpen

Supporting open minds since 2005

Wavelet Transform and Complexity
http://dx.doi.org/10.5772/intechopen.77457
Edited by Dumitru Baleanu

Contributors
Andres Navarro Cadavid, Yor Castano, Jorge Orozco, Beatriz Munoz, Peter Dibal, Elizabeth Onwuka, James Agajo, Caroline Alenoghena, Riccardo Bernardini, Farhan A. Al-Enizi, Awad Al-Asmari, Mario Orlando Oliveira, Jose Horacio Reversat, Lucas Alberto Reynoso, Miguel Delgado-Prieto, Hossam Selim, Fernando Pinal Moctezuma, José Trull, Luis Romeral Martinez, Crina Cojocaru

First published in London, United Kingdom, 2019 by IntechOpen
IntechOpen is the global imprint of INTECHOPEN LIMITED, registered in England and Wales, registration number: 11086078, 7th floor, 10 Lower Thames Street, London, EC3R 6AF, United Kingdom
Printed in Croatia

British Library Cataloguing-in-Publication Data
A catalogue record for this book is available from the British Library

Additional hard and PDF copies can be obtained from orders@intechopen.com

Wavelet Transform and Complexity
Edited by Dumitru Baleanu
p. cm.
Print ISBN 978-1-83962-658-6
Online ISBN 978-1-83962-659-3
eBook (PDF) ISBN 978-1-83962-660-9

For EU product safety concerns:
IN TECH d.o.o., Prolaz Marije Krucifikse Kozulić 3, 51000 Rijeka, Croatia, info@intechopen.com or visit our website at intechopen.com.

Meet the editor

Dumitru Baleanu is a professor at the Institute of Space Sciences, Magurele-Bucharest, Romania, and a visiting staff member at the Department of Mathematics, Cankaya University, Ankara, Turkey. His fields of interest include fractional dynamics and its applications in science and engineering, fractional differential equations, discrete mathematics, mathematical physics, soliton theory, Lie symmetry, dynamic systems on time scales, and the wavelet method and its applications.

Contents

Preface

The wavelet transform is a powerful method applied with great success in several disciplines. For example, when we decompose a signal into wavelets rather than frequencies we can obtain a much better resolution in the domain it is transformed into. Alternatively, when we use a specific wavelet transform the signal is transformed into the wavelet domain. The existence of several wavelet families is a great advantage of the method; however, the question of which is the best family to treat a given data still remains.

This book is structured into two main parts: Analysis and Real-World Applications.

Chapters in Part 1 include: "Using Wavelets for Gait and Arm Swing Analysis" by Yor Jaggy Castaño-Pino, Andrés Navarro, Beatriz Muñoz, and Jorge Luis Orozco; "Analysis of Wavelet Transform Design via Filter Bank Technique" by Peter Yusuf Dibal, Elizabeth Onwuka, James Agajo, and Caroline Alenoghena; and "Wavelets for Differential Equations and Numerical Operator Calculus" by Riccardo Bernardini.

Chapters in Part 2 include: "DWT-Based Data Hiding Technique for Videos Ownership Protection" by Farhan A. Alenizi and Awad Kh. Al-Asmari; "Wavelet Transform Analysis to Applications in Electric Power Systems" by Mario O. Oliveira, J. H. Reversat, and Lucas A. Reynoso; and "Wavelet Transform Applied to Internal Defect Detection by Means of LASER Ultrasound" by Hossam Selim, Fernando Piñal Moctezuma, Miguel Delgado Prieto, José Francisco Trull, Luis Romeral Martínez, and Crina Cojocaru.

The book is designed for students, postdocs, and researchers willing to enter or to go further into the fascinating world of wavelet theory and its applications.

Dumitru Baleanu
Professor,
Institute of Space Sciences,
Magurele, Bucharest, Romania

Department of Mathematics,
Faculty of Arts and Sciences,
Ankara, Turkey

Section 1

Analysis

Chapter 1

Using Wavelets for Gait and Arm Swing Analysis

Yor Jaggy Castaño-Pino, Andrés Navarro, Beatriz Muñoz and Jorge Luis Orozco

Abstract

The human walking pattern can be affected by different factors such as accidents, transplants, or diseases, like Parkinson's disease, which affects motor and mental functions. In motor terms, this disease can generate alterations such as tremors, festination, rigidity, unbalance, slowness, and freezing of gait. Additionally, it is estimated that for the year 2040, the number of people with Parkinson's in the world will be between 12.9 and 14.2 million people. These alarming figures make Parkinson's disease an important focus of attention. In this chapter, we present contributions that suggest wavelet techniques as a useful tool to perform a gait and arm swing analysis; this represents an important approximation that can contribute to describe and differentiate people with Parkinson's disease in early stages of the disease.

Keywords: wavelet, gait analysis, arm swing, Parkinson diagnose, spatiotemporal variables

1. Introduction

Aging is associated with numerous physiological problems that affect the brain. Some of these problems occur in the context of aging, such as cognitive deterioration and motor involvement, and often have an important impact on the central nervous system [1]. The causes of these deficits can be multifactorial and involve the central nervous system, the sensory receptors, the muscles, and the peripheral nerves [2]. On the other hand, there are comorbidities such as Parkinson's disease that can generate an even more marked deterioration of the motor skills of the affected elderly.

Parkinson's disease (PD) is a neurodegenerative disease that mainly affects people older than 60 years and is characterized by a neuronal loss in several areas and brain nuclei, but particularly in the substantia nigra, which can lead initially to motor alterations and delayed cognitive disorders that condition the patient to present physical dependence toward the caregiver and commitment to their autonomy [3].

Among the alterations mentioned are those associated with walking and arm swing. The march and its spatiotemporal characteristics have been analyzed since the Renaissance, and currently the analysis of this has become a very useful tool in the diagnostic evaluation and the severity of the disease, the response to treatment, as well as the impact of therapeutic interventions which can additionally predict the risk of falls [4]. Quantitative gait studies have usually focused on the characteristics of each participant and on the average of steps ignoring the step-by-step fluctuations between subjects. However, it has been shown for two decades that the magnitude of the step-by-step fluctuations and the changes over time during the march (gait dynamics) can be useful to understand the motor control of gait, in the quantification of the pathological and age-related alterations in the locomotion system and in an increase in objectivity in the measurement of mobility and functional status [5].

Motor alterations are one of the key points in the diagnosis of PD patients even in the early stages of the disease; however, the evaluation of the gait may be inconclusive because the slow and short steps are nonspecific and may be related to age, depressive disorders, and other conditions. On the other hand, we must remember that when patients meet the motor criteria for the diagnosis of PD, approximately 70% of the neurons of the substantia nigra have degenerated and the concentration of striatal dopamine has been reduced by 80% [6]; this shows that the typical motor manifestations of PE appear when there is already advanced neurodegeneration, and it has been determined that there is a "preclinical or prediagnostic phase" of PD [7].

Additionally, it is known that PD in early stages can start asymmetrically, since it can affect extremities of a hemibody predominantly and can even differentially affect the upper and lower limbs [8–10]. Thus, the asymmetry in the swing of the arms can be an opportunity for the earlier diagnosis of PE, even in the "prediagnosis stage" [10–14]. The function of swinging the arms during walking is to minimize the angular momentum of the body around the vertical axis [15]; however, there is still controversy as to whether it plays a role in gait stability. The coordination of the lower and upper limbs in the march is a complex phenomenon that has not yet been fully elucidated and involves circuits that we do not yet know. Previously it was thought that the movement of the arms was only passive (like a pendulum due to inertia) and did not imply muscle contraction [16]; however, Braune and Fischer when analyzing the march in a study postulated that this movement should present some muscle activity [17]. Much later, Ballesteros et al. were able to demonstrate with surface electromyography that there is an active muscle component involved which implies some control exerted by superior neural structures [18]. Another study showed that the amplitude of the swing of the arms is partly mediated by muscular activity, since, by reducing it, the amplitude of the swing of the arms decreases markedly by just depending on the passive component [15]. All of the above shows that the arm swing does depend in part on the CNS and can be measured, for example, in PD to observe alterations that correlate with the presence or absence of the disease.

Currently, the main reason for the disability in the world is adjudicated to neurological disorders; one of these is PD, which is the fastest growing, even faster than Alzheimer. In the last 25 years, the prevalence of PD is more than double, which generates double disability and deaths. The Global Burden of Disease study affirms that approximately 6.2 million people have PD. Currently, different subjective tools to assess and diagnose the PD are used in the clinical context; some of these tools are the DGI and UPDRS [19, 20].

However, with the rise of recent technologies, it has become possible to develop tools that allow taking objective measures to complement the diagnosis of Parkinson's; these measures focus on quantifying symptoms of the disease such as tremor, the amount and speed of the steps, as well as the amount of movement in the arms and their speed. However, these tools are considered difficult to access, according to their technological requirements, since they usually require up to 10 specialized cameras, a minimum space of 10 m^2, and must be handled by a clinical expert. Some of the assessment tools used in laboratory settings are motion-capture systems, such as GAITRite, Optitrack, Qualisys, and Vicon. These are used to obtain a quantitative and accurate gait representation, to help the analysis performed by the clinical expert in sport and physical rehabilitation and in gaming industries [20–22]. These systems are characterized by their high cost and complexity, since it requires a minimum technical expertise, enough space to capture test, and a patient preparation and demands a long examination time.

With the technological advance, different motion capture systems of medium complexity have been introduced to the market, able to generate clinically useful variables in medical environments, with a low cost and setup times. Microsoft Kinect was tested for use in the clinical context, as the primary motion-capture device; additionally, it has demonstrated sufficient accuracy for PD assessment through gait analysis [23–29].

In addition to these devices, recent research has focused on efforts to build systems that support the clinical assessment from different perspectives; some of these are force platforms or pressure sensors, which are a set of sensors interconnected and located on the ground along a march corridor, and instrumented shoes, which include small force sensors placed on the template, which are used to detect the initial phases of the march, moments of festination, and freezing of the march.

Other alternatives to tools based on RGBD cameras are those based on wearable devices such as accelerometers or inertial units. With these devices, solutions have been implemented to evaluate movement in the upper extremities and generate metrics to quantify the alterations. Ref. [30] is presented with a system for monitoring and measuring the swing of arms for patients. With Parkinson's disease, this system is composed of a set of handles with accelerometers, which allow extracting variables from the signals that may be related to the alterations generated by the disease, such as the asymmetry in rolling.

2. Wavelet in biomedical applications

For decades, signal processing has been applied to multiple sectors such as industrial, military, health, and entertainment, among others. Regarding the health sector, these techniques have been used to facilitate access to technology and support or complement the diagnosis of a wide variety of diseases. As presented by Suk and Kojima, who use signal processing techniques to clean and extract information from speech signals to make speech recognition, with the purpose of generating a tool that allows disabled people to control by multiple voice home appliances and allow voice control of a wheelchair [31].

Conventional processing techniques and methods allow to filter signals in a frequency range, extract relevant characteristics such as maximum and minimum peaks, fill data by interpolation, and transform and decompose signals in other domains such as frequency and time. Among these processing techniques, wavelet has shown to have a broad application panorama; the literature documents wavelet uses in different and varied fields such as detection of anomalies associated with seismic events in ultralow-frequency geomagnetic signals [32]; it is also possible to use wavelet techniques for image compression, as detailed in [33], who decompose into singular values and use a discrete wavelet transform to improve the maximum ratio of signal-to-noise ratio compared to techniques such as JPEG2000.

In biomedical signals, wavelet transforms have also been suggested for signal compression [34], cardiac pattern recognition [35], EMG classification and decodification [34, 36], feature detection and extraction for ECGs [35] and PPGs [37], and epilepsy diagnosis [38]. Finally, in this chapter, we detail two potential usage scenarios for wavelet techniques, such as gait analysis and arm swing analysis. These two approaches were designed and tested in Parkinson's disease patients, but we consider are not limited to this population, other potential use cases are gait and arm swing alterations in stroke patients, gait analysis in patients with knee replacement, and gait detection and recognition for surveillance.

3. Wavelet background

In this chapter, we apply wavelet decomposition using multiple wavelet mothers, like Daubechies. The discrete wavelet transform (DWT) uses a set of basic functions to perform a decomposition over a $x(n)$ signal in two resultant signals: detailed and approximated signals. The first one is the scaling function, called the basic dilation function. The second one is the main wavelet function. This decomposition is defined by the equation used in [39, 40] and represented as follows:

$$x(n) = \sum_{j=1}^{J} \sum_{k \epsilon Z} d_{2^j}(k)\, \psi^{*}_{J,k}(n) + \sum_{k \epsilon Z} a_{2^j}(k)\, \phi_{J,k}(n) \tag{1}$$

where (1)j is the scale that represents the dilation index and k represents the index in time. J is the decomposition level and $*$ denotes complex conjugation. The wavelet and scaling functions are defined as

$$\phi_{J,k}(n) = 2^{-j/2}\phi(2^{-j}n - k) \tag{2}$$

$$\psi_{J,k}(n) = 2^{-j/2}\psi(2^{-j}n - k) \tag{3}$$

In $\phi_{J,k}(n)$ and $\psi_{J,k}(n)$, j allows the scaling and the wavelet function the dilation or compression. k controls the translation in time. The functions $\phi_{J,k}(n)$ and $\psi_{J,k}(n)$ have the essential properties of low-pass and band-pass Fourier transform, respectively.

The approximation obtained with $a_{2^0}(n)$ at scale $j = 0$ is equivalent to the original signal $x(n)$. The signal $a_{2^j}(n)$ at lower resolutions represents smoothed $a_{2^{j-1}}(k)$. The detailed signals $d_{2^j}(n)$ are given by the difference between approximate signals $a_{2^j}(n)$ and $a_{2^{j-1}}(k)$. The approximate signals $a_{2^j}(n)$ and the detailed signals $d_{2^j}(n)$ are replaced by the following equations:

$$a_{2^j}(n) = \sum_{k} h(k - 2^j n)\, a_{2^{j-1}}(k) \tag{4}$$

$$d_{2^j}(n) = \sum_{k} g(k - 2^j n)\, a_{2^{j-1}}(k) \tag{5}$$

where h and g represent the coefficients of the discrete low-pass and high-pass filters associated with the scaling function and the wavelet function, respectively. Given that each level of wavelet decomposition generates coefficients of length less than the original signal, it is important to clarify that for the use of the approximation and detail coefficients, it was necessary to perform an interpolation process to adjust the size of the coefficients according to the size of the original signal.

4. Capture device

Based on criteria provided by clinical experts, the space selected to record the gait signals with the Microsoft Kinect was a corridor of 1.5 m wide by 4 m long. Each volunteer did walk in the selected space three times. Kinect's represent the joints in a basic human shape with 20 points, three of these points were used (the ankle, the wrist, and the spine base) because they are in the same positions as in the standard anthropometric model used in the benchmark data [24, 26].

To obtain the distance between the Kinect and the subject, we use our eMotion Capture software, which provides the distances to each joint in meters. In the preliminary review [26], we obtain results that suggest the ankle trajectory accurate for gait tracking. The clinical space settings are shown in **Figure 1**. The acceptable capture area was restricted to a distance of 1.5–3.5 m from the camera, which was able to record at least one full gait cycle during each walking test.

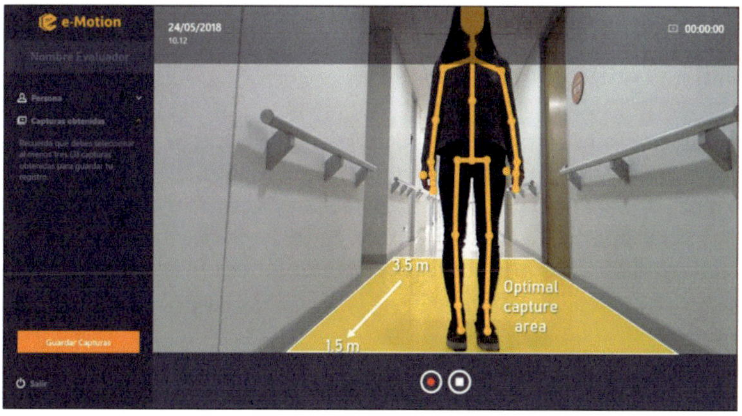

Figure 1.
Graphic interface from eMotion Capture software and acceptable capture area.

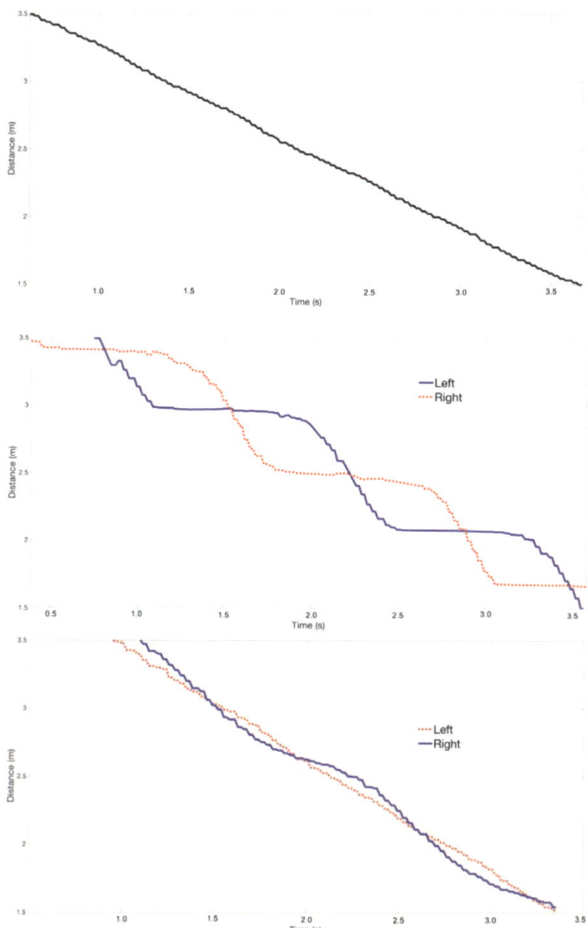

Figure 2.
Signals obtained from Kinect. The first image shows the spine base movement, the second shows the movement related to the left and right ankles, and the third shows the movement related to the left and right wrist.

This software allows us to obtain a representation of the distance between the person and the Kinect, for each articulation of interest, at each instant of time. **Figure 2** shows a representation of the movement of the base of the spine, ankle left and right, and wrist left and right, respectively.

5. Gait analysis with wavelet

In this section, we present the results obtained to apply wavelet in gait signals, obtained with the eMotion Capture system. In this analysis, only the ankle data (left and right) were considered. To generate spatiotemporal variables, we select the best wavelet performance, which was obtained by a comparison between multiple wavelet decomposition and the clinical expert judgment [41].

5.1 Methodology and data

For this study 12 volunteers were selected, 6 women and men, with an age range of 53–73. In each gender group, there were three subjects with early stage PD and three healthy with normal walking patterns. Early stage was defined as stage I or II on the Hoehn and Yahr scale. All participants were evaluated under a dopaminergic agonist, i.e., "on" state. All PD subjects were of completely independent mobility and did not require a walking aid.

5.2 Signal processing with wavelet

The wavelet families tested were Biorthogonal, Coiflets, Daubechies, and Symlets; a total of 12 wavelet decompositions were tested for each gait signal. This was realized with the aim to obtain the best wavelet performance and to observe different spectral- and time-domain information.

To evaluate the wavelet performance, we assess each transformation with the clinical expert criteria. Matlab was used as programming and processing tool; in this software wavelet is defined using an identifier (id) and decomposition value. For example, in "db8" the "db" indicates Daubechies family, and the 8 refers to the vanishing moments. For the present study, we test four wavelet families (Daubechies, Coiflet, Symlet, and Biorthogonal), each wavelet transform with different vanishing moments (db3, db4, db5, db6, db7, db8, coif1, coif2, sym2, sym3, bior2.2, and bior 2.4).

The wavelet transformation was applied with one level of decomposition to each individual ankle signal (left and right). We assess the algorithm applying 12 wavelet decomposition, to 12 subjects, to every ankle, walking in the corridor 3 times. Finally, the system was tested with a total of 864 ankle signals.

Each j level of decomposition is obtained by generating j approximation and detail coefficients, which can be associated to a noise-free version of the original signal and to a version of the noise extracted, respectively.

Figure 3 shows one-level decomposition of a gait signal using wavelet; this process generates two signal, an approximated signal to the original and other with the details extracted.

5.3 Gait phases detection

To distinguish the gait phases, we calculated the mean values of each of the 12 wavelet decompositions we applied to the gait signals, using this as a threshold to distinguish the phases. This threshold was defined as the average value of each

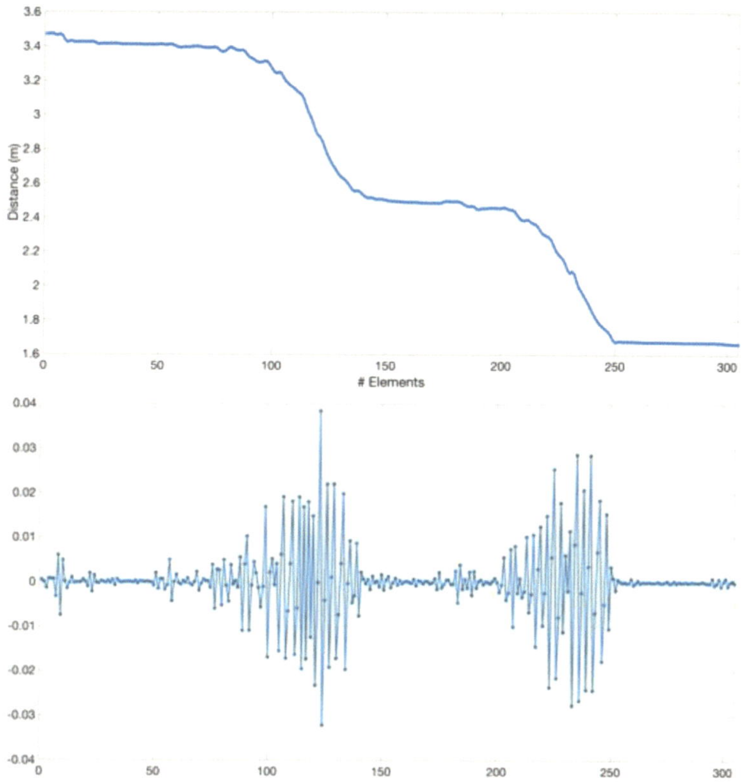

Figure 3.
One-level wavelet decomposition using db8, (a) approximation coefficients, and (b) details coefficients.

wavelet decomposition. To extract the support and swing phases from the ankle signals using this threshold, we defined all values above it as the swing phase and all values below it as the support phase. These allow us to obtain a binary signal with true or 1 when on swing phase and false or 0 when on support phase. **Figure 4** shows one gait signal, the threshold applied to the detail coefficients, and the binary signal generated for this one gait signal.

5.4 Gait phase error detection and correction

From step described in section 5.3 (**Figure 3**), we obtain binary signals, some of these with small intermediate phases. According to gait signals obtained, we set as a criterion that each gait phase should have at least 10 binary elements; some small gait phases do not meet this minimum number of elements and were considered errors. These small intermediate phases are generated due to wavelet sensitivity to detect small changes in signals.

To correct these errors, we designed an algorithm to detect the start and end of each phase and correct for abnormal phases; this algorithm was designed based on the criterion for the minimum number of values that could represent a real gait phase.

5.5 Results

We use Hamming distance [42] as the metric to select the best wavelet transform. This metric was used to compare all the binary gait signals to the ideal reference values. With this we could obtain a quantitative value of the wavelet decomposition accuracies.

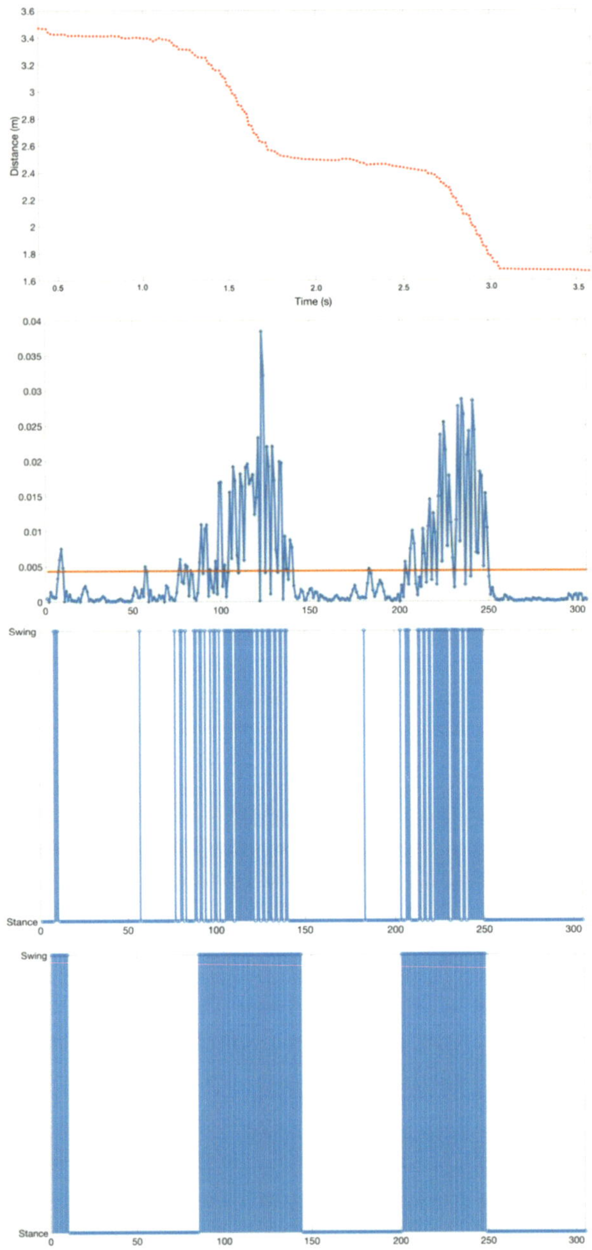

Figure 4.
First image shows the right ankle signal sequence from one subject, who covered about 2 m in about 3 seconds. The second signals show the one-level wavelet decomposition using db8; the red line shows the mean, used as a gait phase classification threshold. The third signal shows the binarized signal, before error correction. The last binary signal shows the ideal gait phase classification, where the gait phases were identified manually by a clinical expert.

The error before and after correction is given in **Table 1**. Before correction the minimum value was 13%, obtained for the db3, db4, db5, bior2.2, and sym3 wavelet members. After correction, the average error was reached for the same wavelet members and by db7 and db8, with 7%. This represents that our algorithm to detect gait phases (stance and swing) has 93% of accuracy, compared with the clinical expert.

After the wavelet comparison, we choose the wavelet "db8" as the member to determine spatiotemporal variables for each subject. Initially, was selected arbitrarily, but later, the "db8" wavelet selection was validated by the statistical comparison.

Wavelet name	Avg error	
	Before correction	**After correction**
Bior2.2	13%	7%
Bior2.4	16%	11%
Coif1	14%	8%
Coif2	17%	11%
Db3	13%	7%
Db4	13%	7%
Db5	13%	7%
Db6	14%	8%
Db7	14%	7%
Db8	14%	7%
Sym2	14%	8%
Sym3	13%	7%

Table 1.
Average error obtained before and after error correction

Variable	Cases		Controls	
	Left	**Right**	**Left**	**Right**
Stance time (s)	2.24 (0.31)	2.17 (0.23)	0.91 (0.10)	1.06 (0.10)
Swing time (s)	1.33 (0.14)	1.33 (0.18)	0.76 (0.09)	0.76 (0.06)
No. of steps	10 (0.55)	9.67 (0.19)	6.83 (0.36)	6.17 (0.29)
Duration time (s)	3.7 (0.41)	3.65 (0.32)	1.72 (0.07)	1.89 (0.09)
Speed test (m/s)	0.63 (0.06)	0.65 (0.05)	1.20 (0.05)	1.04 (0.07)

Table 2.
Average spatiotemporal variable values (standard deviations) obtained for PD and non-PD volunteers

Variable	*p*-Value	
	Left	**Right**
Stance time (s)	0.01	0.04
Swing time (s)	0.02	0.03
No. of steps	0.04	0.03
Duration time (s)	0.01	0.01
Speed test (m/s)	0.01	0.01

Table 3.
p-Values obtained from Mann-Whitney tests

The variables obtained are clinically important and provide objective measures that can be used in the evaluation context to measure and diagnose the PD progression.

The variables presented in **Table 2** are the results obtained for healthy volunteers and PD volunteers. These results suggest significant differences between both groups and represent an objective metric for disease progression quantification. The variables obtained reflect that patients were slower than controls; this is related to the PD gait alterations.

Finally, since PD is an asymmetric disease, we perform a Mann-Whitney test to identify differences statistically significant in the left and right variables for case and control subjects. As shown in **Table 3**, all variables considered provide a mechanism to differentiate PD and non-PD people. The parameters that can be considered as the most appropriate to discriminate patients are stance time, duration time, and test speed.

6. Arm swing analysis with wavelet

In gait analysis with wavelet was important to detect the gait phases; in this case, we were interested in obtaining a measure that allows quantifying the minimum and the maximum displacement of each wrist. For this reason, to generate spatio-temporal variables, we use multiple denoising methods that allow us to obtain a signal without big fluctuations; according to this, we use methods like Savitzky-Golay filter and wavelet decomposition [43]. In this chapter, we present the results obtained of applied wavelet decomposition using db8 to wrist signals.

6.1 Methodology and data

For this study, 25 patients (aged 45–87 years) and 25 controls (aged 46–88 years) were selected, and like in the gait analysis, PD patients were in an early stage of the disease. All participants with PD were under a dopaminergic agonist and were evaluated while in the "on" state. The absence of dementia and any other related to neurological conditions that affect gait was confirmed by an expert neurologist. All PD subjects were completely independent mobility and did not require a walking aid.

6.2 Noise reduction using wavelet

Since the original signals had fluctuations that could affect the analysis and processing, it was necessary to apply wavelet techniques to remove alterations and clean the signal. As showed in **Figure 5**, we apply three-level wavelet decomposition using Daubechies wavelet with eight vanishing moments. From this step, the approximation coefficients at level 3 were used as clean signal.

As a result of the wavelet decomposition, we obtain a clean signal to determine the relative displacement of the wrist, which allows to observe conditions such as rigidity and asymmetry in upper limbs. For the next step, we use the a_3 signal.

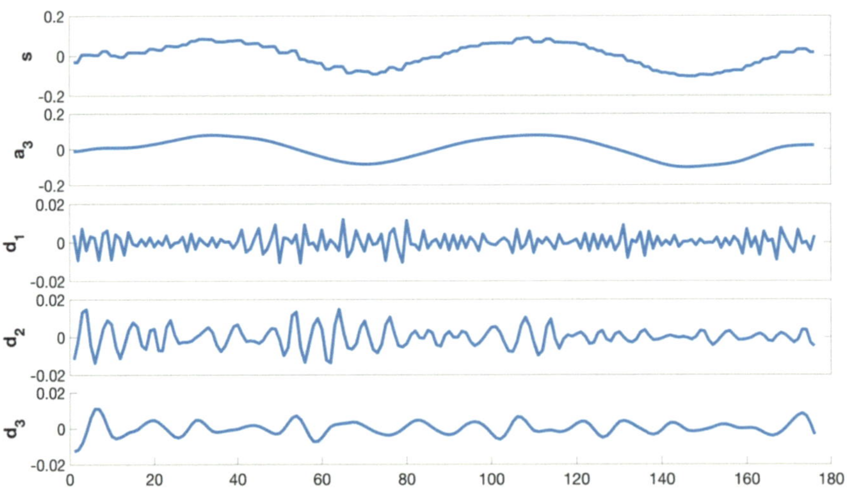

Figure 5.
Approximation coefficient and detail coefficient for wrist signal, the sum of these coefficient level generates original signals ($s = a_3 + d_1 + d_2 + d_3$).

DOI: http://dx.doi.org/10.5772/intechopen.84962

6.3 Swing variables

The arm swing variables calculated using the signal provided by eMotion were arm swing magnitude, arm swing time, arm swing speed, and arm swing asymmetry; these variables are defined as follows:

- Arm swing magnitude: the average distance traveled by the wrist in the anterior/posterior plane, normalized accord the hip center joint [10]

- Arm swing time: duration that took the displacement of a wrist, during a swing cycle, in the anterior/posterior plane

- Arm swing speed: the ratio between the arm swing magnitude and the arm swing time

- Arm swing asymmetry (ASA): proposed by Zifchock et al. and used by Lewek et al. [44], is the outcome of the next equation:

- $$ASA = \frac{\left[45^{\circ} - arct\left(\frac{ArmSwing_{more}}{ArmSwing_{less}}\right)\right]}{90^{\circ}} x\ 100\%$$

6.4 Results

Table 4 shows a comparison of arm swing variables obtained for each limb with the eMotion. This shows that arm swing magnitude (left $p = 0.002$, right $p = 0.006$) and arm swing speed (left $p = 0.002$, right $p = 0.004$) were significantly reduced in the PD group for both limbs. The control group shows a lowest arm swing asymmetry than the patient group ($p < 0.001$). Based on the side, the variables that show significant differences for the left side were arm swing magnitude, speed, and ASA and for the right side were arm swing magnitude, speed, and ASA. Also, the most affected side determined with Kinect and the one with the highest score of the pondered items of the MDS-UPDRS-III were compared. These comparisons suggest that our device is to recognize the most affected side in the 80% of cases. Due to the limited sample size, differences in the symmetrical group were not evaluated.

Arm swing variables	Left wrist (n:50)		p-Value (left wrist)	Right wrist (n:50)		p-Value (right wrist)
	PD patients	Healthy subjects		PD patients	Healthy subjects	
Arm swing magnitude	0.16 (IQR 0.08–0.2)	0.26 (IQR 0.17–0.33)	0.002	0.16 (IQR 0.09–0.24)	0.26 (IQR 0.20–0.34)	0.006
Arm swing time	0.99 (IQR 0.93–1.12)	1.09 (IQR 0.94–1.15)	0.171	0.98 (IQR 0.90–1.03)	1.05 (IQR 0.96–1.12)	0.177
Arm swing speed	0.16 (IQR 0.08–0.2)	0.25 (IQR 0.18–0.29)	0.002	0.14 (IQR 0.09–0.21)	0.26 (IQR 0.18–0.31)	0.004
	PD patients			Healthy subjects		p-Value
ASA	0.16 (IQR 0.09–0.23)			0.063 (IQR 0.03–0.08)		<0.001

Table 4.
Arm swing differences between PD patients and the healthy subject group.

7. Limitations

The use of Kinect® in this clinical context has reported relative and overall reliability regarding spatiotemporal parameters [45–47]. Further advances in software and hardware are essential to further enhance Kinect's® sensitivity for kinematic measurements [48, 49]. Nevertheless, since the RGBD cameras, like Kinect, are low-cost and portable devices, this represents an opportunity in the field of telemedicine, allowing easy access to gait assessment in the clinical space and allowing remote diagnose in rural areas, where there are no clinical experts. Finally, it is remarkable how the eMotion Capture system can calculate and automatically obtain the gait cycle variables that are considered relevant for decision-making processes in the clinical context of patients with Parkinson's disease.

Some authors propose that arm swing analysis could help in the differentiation between TD and PIG subtypes, but the small sample in this study limited the subgroup analysis. Because Kinect was discontinued from the market, in future research jobs, alternative RGBD cameras, such as the Intel RealSense D435, needs to be investigated.

8. Conclusion

From the previous sections, the use of wavelet techniques for gait analysis and arm swing was detailed. Finally, we can conclude that it is possible to use wavelet techniques to automate and quantify spatiotemporal variables related to the gait, to perform an objective analysis of Parkinson's disease. In addition to this, the eMotion system has demonstrated to be a useful tool that can be used in a clinical context to generate spatiotemporal variables like arm swing asymmetry, arm swing speed, swing magnitude, stance time, swing time, step number, duration test, and speed, which are useful for differentiating PD patients from healthy individuals.

Acknowledgements

We thank our colleague Jaime Valderrama and Juan David Arango for the support and time invested in the current research job.

Author details

Yor Jaggy Castaño-Pino[1], Andrés Navarro[1]*, Beatriz Muñoz[2] and Jorge Luis Orozco[2]

1 i2t Research Team, Icesi University, Cali, Colombia

2 Fundación Valle del Lili, Cali, Colombia

*Address all correspondence to: anavarro@icesi.edu.co

References

[1] Stranahan AM, Mattson MP. Recruiting adaptive cellular stress responses for successful brain ageing. Nature Reviews. Neuroscience. 2012;**13**:209-216

[2] Seidler RD, Bernard JA, Burutolu TB, et al. Motor control and aging: Links to age-related brain structural, functional, and biochemical effects. Neuroscience and Biobehavioral Reviews. 2010;**34**:721-733

[3] Fernandez HH. 2015 update on Parkinson disease. Cleveland Clinic Journal of Medicine. 2015;**82**:563-568

[4] Hoskovcová M, Dušek P, Sieger T, et al. Predicting falls in Parkinson disease: What is the value of instrumented testing in OFF medication state? PLoS One. 2015;**10**:e0139849

[5] Hausdorff JM. Gait dynamics, fractals and falls: Finding meaning in the stride-to-stride fluctuations of human walking. Human Movement Science. 2007;**26**:555-589

[6] Becker G, Müller A, Braune S, et al. Early diagnosis of Parkinson's disease. Journal of Neurology. 2002;**249**(Suppl 3: III):40-48

[7] Noyce AJ, Lees AJ, Schrag A-E. The prediagnostic phase of Parkinson's disease. Journal of Neurology, Neurosurgery, and Psychiatry. 2016;**87**:871-878

[8] Frazzitta G, Pezzoli G, Bertotti G, et al. Asymmetry and freezing of gait in parkinsonian patients. Journal of Neurology. 2013;**260**:71-76

[9] Maetzler W, Hausdorff JM. Motor signs in the prodromal phase of Parkinson's disease. Movement Disorders: Official Journal of the Movement Disorder Society. 2012;**27**:627-633

[10] Lewek MD, Poole R, Johnson J, et al. Arm swing magnitude and asymmetry during gait in the early stages of Parkinson's disease. Gait & Posture. 2010;**31**:256-260

[11] Dietz V, Fouad K, Bastiaanse CM. Neuronal coordination of arm and leg movements during human locomotion. The European Journal of Neuroscience. 2001;**14**:1906-1914

[12] Mirelman A, Bernad-Elazari H, Thaler A, et al. Arm swing as a potential new prodromal marker of Parkinson's disease. Movement Disorders: Official Journal of the Movement Disorder Society. 2016;**31**:1527-1534

[13] Huang X, Mahoney JM, Lewis MM, et al. Both coordination and symmetry of arm swing are reduced in Parkinson's disease. Gait & Posture. 2012;**35**:373-377

[14] Sterling NW, Cusumano JP, Shaham N, et al. Dopaminergic modulation of arm swing during gait among Parkinson's disease patients. Journal of Parkinson's Disease. 2015;**5**:141-150

[15] Goudriaan M, Jonkers I, van Dieen JH, et al. Arm swing in human walking: What is their drive? Gait & Posture. 2014;**40**:321-326

[16] Meyns P, Bruijn SM, Duysens J. The how and why of arm swing during human walking. Gait & Posture. 2013;**38**:555-562

[17] Braune W, Fischer O. The Human Gait. Berlin Heidelberg: Springer-Verlag; 1987. Available from: www.springer.com/de/book/9783642703287 [Accessed: November 21, 2018]

[18] Ballesteros ML, Buchthal F, Rosenfalck P. The pattern of muscular activity during the arm swing of natural walking. Acta Physiologica Scandinavica. 1965;**63**:296-310

[19] Dorsey ER, Bloem BR. The Parkinson pandemic—A call to action. JAMA Neurology. 2018;**75**:9-10

[20] Movement Disorder Society Task Force on Rating Scales for Parkinson's Disease. The unified Parkinson's disease rating scale (UPDRS): Status and recommendations. Movement Disorders: Official Journal of the Movement Disorder Society. 2003;**18**:738-750

[21] VICON. Vicon Motion Capture Systems. VICON; 2018. Available from: http://www.vicon.com [Accessed: November 21, 2018]

[22] GAITRite. World Leader in Temporospatial Gait Analysis. 2018. Available from: https://www.gaitrite.com [Accessed: November 21, 2018]

[23] Rocha AP, Choupina H, Fernandes JM, et al. Parkinson's disease assessment based on gait analysis using an innovative RGB-D camera system. In: Conf Proc Annu Int Conf IEEE Eng Med Biol Soc; IEEE Eng Med Biol Soc Annu Conf, 2014. 2014. pp. 3126-3129

[24] Microsoft Corporation. Kinect for Windows Sensor Components and Specifications. Available from: https://developer.microsoft.com/en-us/windows/kinect

[25] Cicchetti DV, Sparrow SA. Developing criteria for establishing interrater reliability of specific items: Applications to assessment of adaptive behavior. American Journal of Mental Deficiency. 1981;**86**:127-137

[26] Arango Paredes JD, Muñoz B, Agredo W, et al. A reliability assessment software using Kinect to complement the clinical evaluation of Parkinson's disease. In: Conf Proc Annu Int Conf IEEE Eng Med Biol Soc; IEEE Eng Med Biol Soc Annu Conf; 2015. 2015. pp. 6860-6863

[27] Galna B, Barry G, Jackson D, et al. Accuracy of the Microsoft Kinect sensor for measuring movement in people with Parkinson's disease. Gait & Posture. 2014;**39**:1062-1068

[28] Bloem BR, Marinus J, Almeida Q, et al. Measurement instruments to assess posture, gait, and balance in Parkinson's disease: Critique and recommendations. Movement Disorders: Official Journal of the Movement Disorder Society. 2016;**31**:1342-1355

[29] Ťupa O, Procházka A, Vyšata O, et al. Motion tracking and gait feature estimation for recognising Parkinson's disease using MS Kinect. Biomedical Engineering Online. 2015;**14**:97

[30] Rincón D, Navarro A. Arm swinging measurement and monitor system for patients diagnosed with Parkinson's disease. In: Proceedings of the IV School on Systems and Networks. Valdivia, Chile; 2018. pp. 53-56

[31] Suk S, Kojima H. Voice activated appliances for severely disabled persons. In: Speech Recognition. 2008. DOI: 10.5772/6361

[32] Alegria OC, Valtierra-Rodriguez MP, Amezquita-Sanchez J, et al. Empirical wavelet transform-based detection of anomalies in ULF geomagnetic signals associated to seismic events with a fuzzy logic-based system for automatic diagnosis. In: Wavelet Transform Some Its Real-World Appl. 2015. DOI: 10.5772/61163

[33] Anbarjafari G, Rasti P, Daneshmand M, Ozcinar C. Resolution enhancement based image compression technique using singular value decomposition and wavelet transforms. In: Wavelet Transform Some Its Real-World Appl. 2015. DOI: 10.5772/61335

[34] Gradolewski D, Tojza PM, Jaworski J, et al. Arm EMG wavelet-based denoising system. In: Awrejcewicz J, Szewczyk R, Trojnacki M, et al., editors. Mechatronics—Ideas for Industrial Application. Cham, Switzerland: Springer International Champions; 2015. pp. 289-296

[35] Li C, Zheng C, Tai C. Detection of ECG characteristic points using wavelet

transforms. IEEE Transactions on Biomedical Engineering. 1995;**42**:21-28

[36] Chau T. A review of analytical techniques for gait data. Part 2: Neural network and wavelet methods. Gait & Posture. 2001;**13**:102-120

[37] Cvetkovic D, Ubeyli E, Cosic I. Wavelet transform feature extraction from human PPG, ECG, and EEG signal responses to ELF PEMF exposures: A pilot study. Digital Signal Processing. 2008;**18**:861-874

[38] Akin M, Arserim MA, Kiymik MK, et al. A new approach for diagnosing epilepsy by using wavelet transform and neural networks. In: 2001 Conference Proceedings of the 23rd Annual International Conference of the IEEE Engineering in Medicine and Biology Society. Vol. 2. 2001. pp. 1596-1599

[39] Sekine M, Tamura T, Akay M, et al. Discrimination of walking patterns using wavelet-based fractal analysis. IEEE Transactions on Neural Systems and Rehabilitation Engineering. 2002;**10**:188-196

[40] Akay M. Wavelets in biomedical engineering. Annals of Biomedical Engineering. 1995;**23**:531-542

[41] Muñoz B, Castaño-Pino YJ, Paredes JDA, et al. Automated gait analysis using a Kinect camera and wavelets. In: IEEE 20th International Conference on e-Health Networking, Applications and Services (Healthcom). 2018. pp. 1-5

[42] Md Saad R, Ahmad MZ, Abu MS, et al. Hamming distance method with subjective and objective weights for personnel selection. The Scientific World Journal. 2014. pp.1-9. DOI: 10.1155/2014/865495

[43] Ospina BM, Chaparro JAV, Paredes JDA, et al. Objective arm swing analysis in early-stage Parkinson's disease using an RGB-D camera (Kinect®). Journal of Parkinson's Disease. 2018;**8**:563-570

[44] Zifchock RA, Davis I, Higginson J, et al. The symmetry angle: A novel, robust method of quantifying asymmetry. Gait & Posture. 2008;**27**:622-627

[45] Müller B, Ilg W, Giese MA, et al. Validation of enhanced kinect sensor based motion capturing for gait assessment. PLoS One. 2017;**12**:e0175813

[46] Cunha JPS, Rocha AP, Choupina HMP, et al. A novel portable, low-cost kinect-based system for motion analysis in neurological diseases. In: 2016 38th Annual International Conference of the IEEE Engineering in Medicine and Biology Society (EMBC). 2016. pp. 2339-2342

[47] Eltoukhy M, Kuenze C, Oh J, et al. Microsoft Kinect can distinguish differences in over-ground gait between older persons with and without Parkinson's disease. Medical Engineering & Physics. 2017;**44**:1-7

[48] Hausdorff JM. Gait dynamics in Parkinson's disease: Common and distinct behavior among stride length, gait variability, and fractal-like scaling. Chaos (Woodbury, N.Y.). 2009;**19**:026113

[49] Ko S, Hausdorff JM, Ferrucci L. Age-associated differences in the gait pattern changes of older adults during fast-speed and fatigue conditions: Results from the Baltimore longitudinal study of ageing. Age and Ageing. 2010;**39**:688-694

Chapter 2

Analysis of Wavelet Transform Design via Filter Bank Technique

Peter Yusuf Dibal, Elizabeth Onwuka, James Agajo and Caroline Alenoghena

Abstract

The technique of filter banks has been extensively applied in signal processing in the last three decades. It provides a very efficient way of signal decomposition, characterization, and analysis. It is also the main driving idea in almost all frequency division multiplexing technologies. With the advent of wavelets and subsequent realization of its wide area of application, filter banks became even more important as it has been proven to be the most efficient way a wavelet system can be implemented. In this chapter, we present an analysis of the design of a wavelet transform using the filter bank technique. The analysis covers the different sections which make up a filter bank, i.e., analysis filters and synthesis filters, and also the upsamplers and downsamplers. We also investigate the mathematical properties of wavelets, which make them particularly suitable in the design of wavelets. The chapter then focuses attention to the particular role the analysis and the synthesis filters play in the design of a wavelet transform using filter banks. The precise procedure by which the design of a wavelet using filter banks can be achieved is presented in the last section of this chapter, and it includes the mathematical techniques involved in the design of wavelets.

Keywords: wavelet, filter bank, perfect reconstruction, orthogonality, paraunitary condition

1. Introduction

Filter banks can be defined as the cascaded arrangement of filters, i.e., low-pass, high-pass, and band-pass filters connected by sampling operators in such a manner as to achieve the decomposition and recomposition of a signal from a spectrum perspective. The sampling operators could either be downsamplers or upsamplers. The downsamplers are called decimators while the upsamplers are called expanders. The technique of filter banks plays an important role in most digital systems that rely on signal processing for their operations. Using this technique, any signal feature can be reliably extracted and analyzed; hence filter banks have wide applications in digital signal processing systems. A filter bank as shown in **Figure 1** [1, 2] consists of different parts, which collectively execute a desired function.

As can be seen in **Figure 1**, the filter bank is made of two sections: the analysis filter bank section (composed of analysis filters and downsamplers), and the synthesis filter bank section (composed of upsamplers and synthesis filters). In this chapter, we will discuss the analysis and synthesis filter bank sections, their

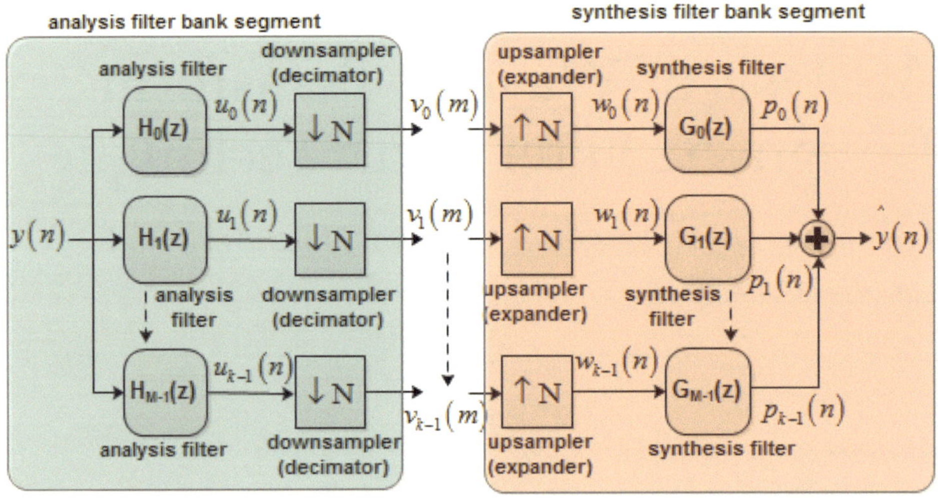

Figure 1.
k-Channel filter bank [1, 2].

responses to incoming signals, and how they work together in the derivation of a wavelet transform function.

2. Analysis filter bank section

The analysis filter bank section is made up of the analysis filter banks, and downsamplers or decimators which together act on an input signal to perform a desired function through decomposition of the signal. In this section, we will analyze the mathematical relationship that exists between these two components. To have a thorough understanding of this relationship, it is important to briefly discuss these components separately.

2.1 Analysis filter bank

The filters that make up the analysis filter banks could either be low-pass filters, or high-pass filters. Each of these filters, as shown in **Figure 2**, allows the passage of only a particular frequency component of the input signal $y(n)$. Thus, specific features of the input signal embedded at different frequencies can be individually extracted and investigated using the analysis filter bank [3, 4]. The k-channel filter bank in **Figure 2** separates the frequencies of the input signal in the manner presented.

It can be seen from the frequency responses that the output of the filters overlap each other. This is because in practice, the filters are not ideal. However, the overlapping condition can be improved through an optimized design of the filters. Mathematically, the effect of each of the filters in the filter bank on the input signal $y(n)$ can be stated as follows:

$$\begin{aligned} U_0(Z) &= Y(Z)H_0(Z) \\ U_1(Z) &= Y(Z)H_1(Z) \\ U_2(Z) &= Y(Z)H_2(Z) \\ U_{k-1}(Z) &= Y(Z)H_{M-1}(Z) \end{aligned} \tag{1}$$

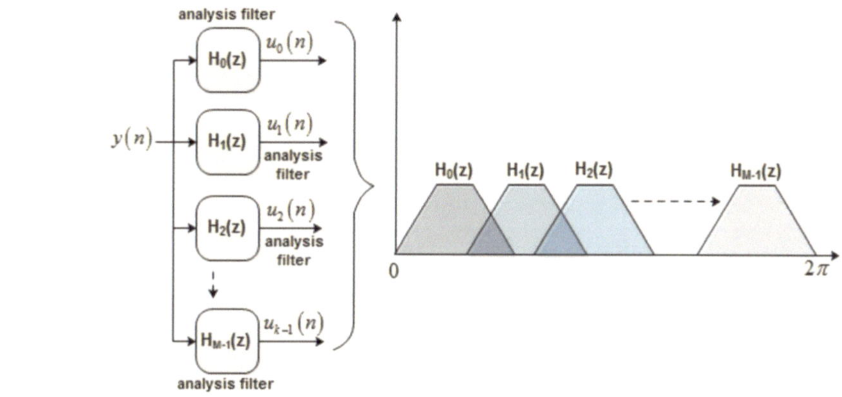

Figure 2.
Separation of input signals into sub-band frequencies by analysis filter bank.

where $U_i(z)$ is the z-transform of the result from the convolution operation between the z-transform of the input signal $Y(Z)$ and the z-transform of the filter $H_i(Z)$. The output $U_i(z)$ in **Figure 2** is fed into the corresponding downsampler of **Figure 1**. In the next section, we will analyze the downsampler and state the mathematical operation it performs on a given signal.

2.2 Downsampler/decimator

The downsampler shown in **Figure 1** downsamples an input signal by a factor of N. This implies that it only retains all the Nth samples in a given sequence. For example, if $N = 2$, then the downsampler will retain all even samples in a given sequence. Given an input signal $x(n)$, the downsampler with a factor of 2 will downsample the signal as:

$$\hat{x}(n) = x(2n), \forall n \in \mathbb{Z} \tag{2}$$

Figure 3 shows the conceptual depiction of the relationship in Eq. (2).

Mathematically, the output of the decimator in **Figure 1** can be expressed as a product of the input sequence $u_i(n)$ and the sequence of unit impulses which are N samples apart, i.e.,

$$v_i(n) = \sum_{k \in \mathbb{Z}} u_i(n)\delta(n - kN), \forall k \in \mathbb{Z} \tag{3}$$

The relationship in Eq. (3) will only select the kN^{th} sample of $u_i(n)$, and the Fourier series expansion of the impulse series can be expressed as [5]:

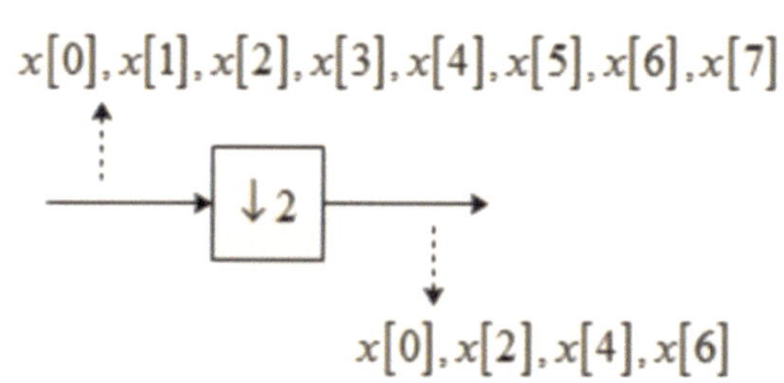

Figure 3.
Decimation by a factor of 2.

$$\sum_{k \in \mathbb{Z}} \delta(n - kN) = \frac{1}{N} \sum_{k=0}^{N-1} e^{-j2\pi kn/N} \tag{4}$$

Setting $W_N = e^{-j2\pi/N}$ and $n = 1$, the relationship in Eq. (4) becomes:

$$\sum_{k \in \mathbb{Z}} \delta(n - kN) = \frac{1}{N} \sum_{k=0}^{N-1} W_N^{-k} \tag{5}$$

Substituting Eq. (5) into (3) yields:

$$v_i(n) = \frac{1}{N} \sum_{k=0}^{N-1} u_i(n) W_N^{-k} \tag{6}$$

In terms of z-transformation, the relationship in Eq. (6) can be expressed as:

$$V_i(Z) = \frac{1}{N} \sum_{k=0}^{N-1} U_i\left(Z^{\frac{1}{N}} W_N^{-k}\right) \tag{7}$$

Having looked at the analysis filters and downsamplers, we will now turn our attention to synthesis filter bank section of **Figure 1**.

3. Synthesis filter bank section

The synthesis filter bank section is made of the upsamplers and synthesis filter banks. These components work together to perform the opposite operation performed by the analysis filter bank section shown in **Figure 1**. In this section, we will make an analysis of the mathematical relationship that governs the operation of the synthesis filters and upsamplers.

3.1 Synthesis filter bank

Similar to the analysis filter bank, the synthesis filter bank is made of low-pass and high-pass filters. The output of these filters as shown in **Figure 1**, are summed to a common output. In typical filter bank applications, the frequency responses of these filters are typically matched to those of the analysis filters shown in **Figure 2**. The mathematical expression for the effect each of these filters has on the corresponding input signal $w_i(n)$ is as stated below [6]:

$$\begin{aligned} P_0(Z) &= W_0(Z)G_0(Z) \\ P_1(Z) &= W_1(Z)G_1(Z) \\ P_2(Z) &= W_2(Z)G_2(Z) \\ P_{k-1}(Z) &= W_{k-1}(Z)G_{M-1}(Z) \end{aligned} \tag{8}$$

In **Figure 2**, the input to the synthesis filter bank is upsamplers or expanders. The next section gives a brief review of the upsamplers.

3.2 Upsampler/expander

The upsampler expands an input signal by a factor N. It does this by inserting zeros at every nth position in the sequence of the input signal. For example, if

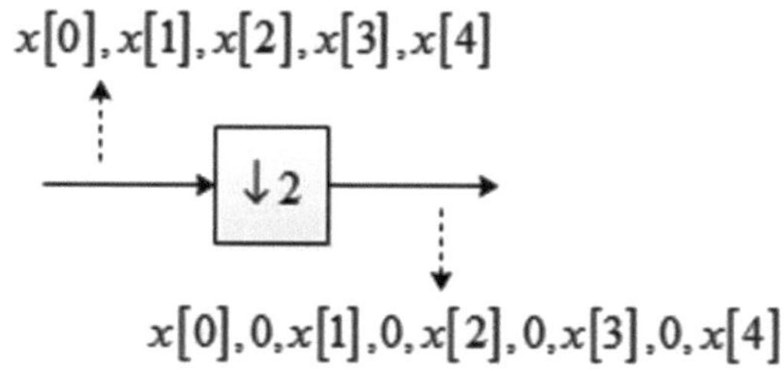

Figure 4.
Upsampling by a factor of 2.

$N = 2$, then the upsampler will insert a zero between every two adjacent samples in a given sequence as shown in **Figure 4**.

Given an input signal $v_i(n)$ in **Figure 1**, an upsampler with a factor of 2 will upsample the signal using the relationship [7]:

$$w_i(n) = \sum_{k \in \mathbb{Z}} v_i(n)\delta(n - kN), \forall k \in \mathbb{Z} \tag{9}$$

Similar to the expression in Eq. (3), the z-transform of the expression in Eq. (9) which is an upsampler is stated as follows [8]:

$$W_i(Z) = \frac{1}{N}\sum_{k=0}^{N-1} V_i\left(Z^N W_N^{-k}\right) \tag{10}$$

To be useful in wavelet designs, filter banks must be designed to have certain characteristics which guarantee that a signal at the input of a filter bank will be received accurately at the output of the filter bank. In the next section, we will examine the properties of filter banks and how these properties influence the design of wavelets.

4. Properties of filter banks for wavelet design

In wavelet designs, filter banks are required to possess three important properties which are fundamental to the realization of a wavelet function. These properties include: perfect reconstruction, orthogonality, and paraunitary condition.

4.1 Perfect reconstruction

This property guarantees that the signal at the output of a given filter bank is a delayed version of the signal at the input of the filter bank. Perfect reconstruction is an important property of a filter bank because it cancels the effect of aliasing of the input signal at the output, caused by the downsamplers and upsamplers. To understand this point, consider a two-channel finite impulse response FIR filter bank shown in **Figure 5**.

The output $\hat{y}(n)$ is derived using Eqs. (6) and (10) as follows in terms of the signal component and aliasing component as:

$$\hat{Y}(z) = \text{signal\_component} + \text{aliasing\_component} \tag{11}$$

where the signal_component and aliasing_component are defined as:

$$\left.\begin{aligned} \text{signal\_component} &= \frac{1}{2}[F_0(z)H_0(z) + F_1(z)H_1(z)]X(z) \\ \text{aliasing\_component} &= \frac{1}{2}[F_0(z)H_0(-z) + F_1(z)H_1(-z)]X(-z) \end{aligned}\right\} \tag{12}$$

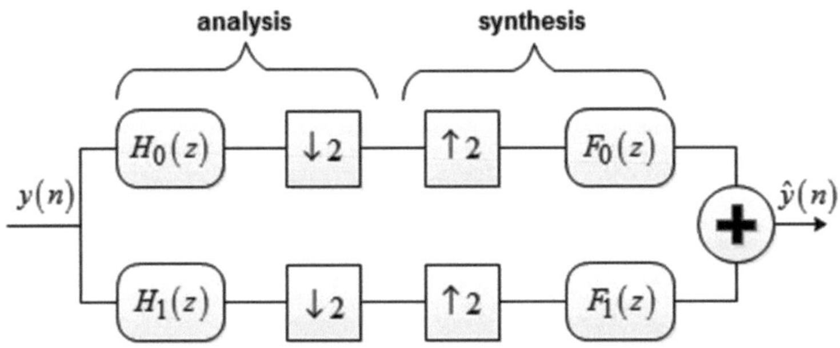

Figure 5.
Two-channel FIR filter bank.

To achieve perfect reconstruction, the following condition must be satisfied [1]:

$$\left.\begin{aligned} F_0(z)H_0(z) + F_1(z)H_1(z) &= 2z^{-1} \\ F_0(z)H_0(-z) + F_1(z)H_1(-z) &= 0 \end{aligned}\right\} \tag{13}$$

The relationships in Eqs. (11) and (13) are possible when the filter bank is constructed as a QMF (quadrature mirror filter) filter bank or CQF (conjugate quadrature filter) filter bank. Both QMF and CQF banks provide a mechanism by which complete cancellation of the aliasing component in Eq. (11) can be accomplished. Using QMF, aliasing cancellation can be achieved by constructing the filters in **Figure 5** based on the following relationships [4, 5]:

$$\left.\begin{aligned} F_0(z) &= H_0(z) \\ H_1(z) &= H_0(-z) \\ F_1(z) &= -H_1(z) \end{aligned}\right\} \tag{14}$$

In Eq. (14), the synthesis filter $F_0(z)$ has the same coefficients as the analysis filter $H_0(z)$; the analysis filter $H_1(z)$ has the same coefficients as the analysis filter $H_0(z)$, but every other value is negated; the synthesis filter $F_1(z)$ is a negative copy of the analysis filter $H_1(z)$. For example, if the analysis filter $H_0(z)$ has coefficientsp, q, r, s, then the filter bank in **Figure 5** will assume the structure shown in **Figure 6**.

For the CQF bank, the coefficients of the analysis filter $H_1(z)$ are a reversed version of the analysis filter $H_0(z)$ with every other value negated. The synthesis filters $F_0(z)$ and $F_1(z)$ are a reversed versions of the analysis filters $H_0(z)$ and $H_1(z)$, respectively. These relationships can be stated mathematically as follows [10]:

$$\left.\begin{aligned} H_1(z) &= z^{-1}H_0(-z^{-1}) \\ F_0(z) &= H_1(-z) \\ F_1(z) &= -H_0(-z) \end{aligned}\right\} \tag{15}$$

Based on the relationship in Eq. (15), the filter bank shown in **Figure 6** for CQF will assume the structure shown in **Figure 7**.

Based on the structure of **Figures 6** or 7, the output signal $\hat{y}[n]$ is related to the input signal $y[n]$ by the expression:

$$\hat{y}[n] = (pp + qq + rr + ss)y[n-3] \tag{16}$$

If we impose the condition that $pp + qq + rr + ss = 1$, then Eq. (16) becomes:

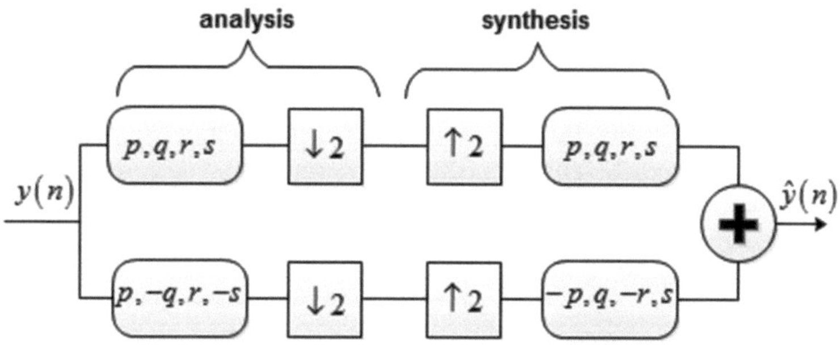

Figure 6.
QMF two-channel FIR filter bank.

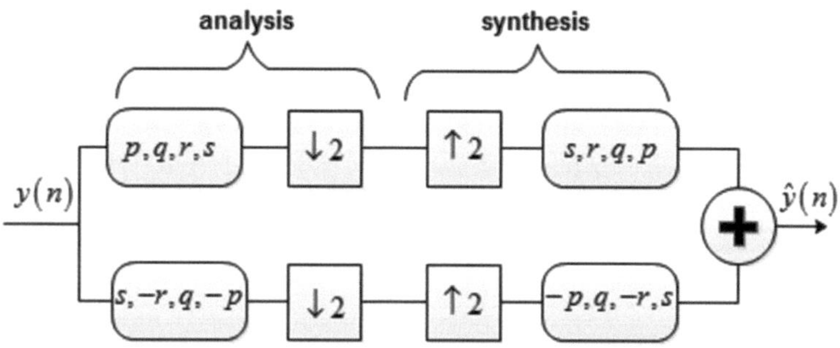

Figure 7.
CQF two-channel FIR filter bank.

$$\hat{y}[n] = y[n-3] \tag{17}$$

The relationship in Eq. (17) states that the output signal $\hat{y}[n]$ is delayed version of the input signal $y[n]$ by three samples. We leave the verification of Eq. (16) as an exercise for the reader.

Having looked at perfect reconstruction as a necessary property for a filter bank in wavelet design, we now look at orthogonality as also an essential property for a filter bank in the design of wavelets.

4.2 Orthogonality

Orthogonality in a filter bank is a situation in which the synthesis filter bank is a transpose of the analysis filter bank. This is a useful property in the sense that it allows for the energy preservation of the signal being processed. This important property is achieved through the imposition of the orthogonality condition on both the analysis and filter bank sections while at the same time preserving the perfect reconstruction condition of the filter bank. The imposition of the orthogonality condition in a filter bank (see **Figure 5**) occurs when the following relationships are satisfied [11]:

$$\left\langle \tilde{f}_0(n-2k), h_1(n-2l) \right\rangle = 0 \left\langle \tilde{f}_1(n-2k), h_0(n-2l) \right\rangle = 0 \tag{18}$$

where

$$\tilde{g}_i(n) = g_i(-n)$$

and

$$\langle \tilde{g}_0(n-2k), h_0(n) \rangle = \delta_k \; \langle \tilde{g}_1(n-2k), h_1(n) \rangle = \delta_k \tag{19}$$

In Eq. (18), the inner product of the coefficients of the synthesis filter $F_0(z)$ and the analysis filter $H_1(z)$ must be zero and the inner product of the coefficients of the synthesis filter $F_1(z)$ and the analysis filter $H_0(z)$ must also be zero for the orthogonality condition to hold.

Also, the low-pass analysis filter $H_0(z)$ is related to the other three filters through the following expressions [12]:

$$\left.\begin{aligned} H_1(z) &= cz^{-(L-1)}\tilde{H}_0(-z) \\ F_0(z) &= z^{-(L-1)}\tilde{H}_0(z) \\ F_1(z) &= z^{-(L-1)}\tilde{H}_1(z) \end{aligned}\right\} \tag{20}$$

where L denotes the length of the filter which must be even, and c is a constant with $|c| = 1$; $\tilde{H}_0(-z)$ is the flipped and conjugated version of $H_0(z)$, $\tilde{H}_0(z)$ is the conjugated version of $H_0(z)$, and $\tilde{H}_1(z)$ is the conjugated version of $H_1(z)$.

The condition in Eq. (20) also describe the necessary requirement for a filter bank to be paraunitary (which we shall examine in the next section), i.e., the low-pass filter $H_0(z)$ satisfy the following power symmetry of halfband condition [8, 9]:

$$P(z) + P(-z) = 2 \tag{21}$$

where $P(z) = H_0(z)\hat{H}_0(z)$. If the low-pass filter $H_0(z)$ satisfies the required symmetry condition:

$$H_0(z) = z^{-(L-1)}H_0\left(z^{-1}\right) \tag{22}$$

then $P(z)$ is said to be a real filter. The implication of the constraint in Eq. (21) is that $H_1(z)$ and $F_1(z)$ be antisymmetric filters, and $F_0(z)$ is a symmetric filter. The relationships in Eqs. (20)–(22) give the necessary and sufficient condition for the characterization of a filter bank with orthogonality and symmetry.

The orthogonality condition for a filter bank can also be examined from a polyphase perspective. Consider the polyphase representation of the filter bank in **Figure 5** as illustrated in **Figure 8** [13].

If $\mathbf{E}(z)$ in **Figure 8** is type-I analysis polyphase matrix, and $\mathbf{R}(z)$ is type-II synthesis polyphase matrix, then [13]:

$$\begin{aligned} [H_0(z) \;\; H_1(z)]^T &= \mathbf{E}(z^2)\left[1 \quad z^{-1}\right]^T \\ \left[F_0(z) \quad F_1(z)\right] &= \left[z^{-1} \quad 1\right]\mathbf{R}(z^2) \end{aligned} \tag{23}$$

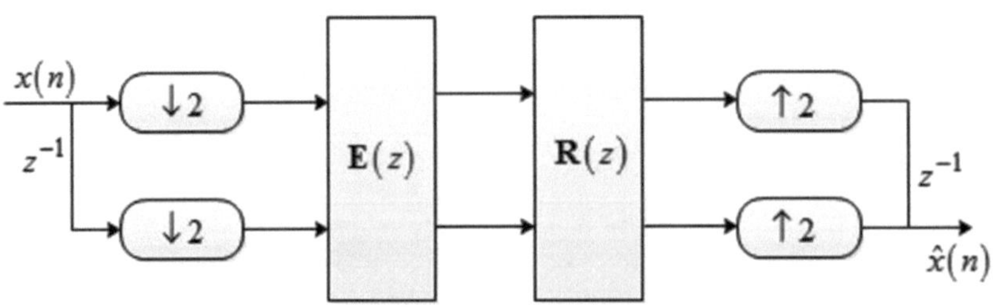

Figure 8.
Polyphase implementation of filter bank.

The conditions in Eqs. (20)–(22) hold true iff $\mathbf{E}(z)$ and $\mathbf{R}(z)$ satisfy the following:

$$\left.\begin{array}{r}\hat{\mathbf{E}}(z)\mathbf{E}(z)=\mathbf{I}\\ \mathbf{R}(z)=z^{-(k-1)}\mathbf{E}(z)\\ \mathbf{E}(z)=z^{-(k-1)}\mathrm{diag}(1,-1)\mathbf{E}(z^{-1})\mathbf{J}\end{array}\right\} \tag{24}$$

where $k=L/2$, with the first and second condition in Eq. (24) relating to the filter bank orthogonality condition, and the last represents the filter bank symmetry.

We now look at the paraunitary condition of a filter bank, which is also a necessary property in filter bank implementation of wavelets.

4.3 Paraunitary condition

In the filter bank implementation of a wavelet transform, the paraunitary condition plays the critical role of guaranteeing the generation of orthonormal wavelets, and also perfect recovery of a decomposed signal. The paraunitary condition guarantees that recovered signal will suffer no phase or aliasing effect if a filter bank satisfies the paraunitary condition [14].

Given a polyphase transfer function matrix $\mathbf{E}(z)$, the paraunitary condition is established by the matrix iff [15]:

$$\mathbf{E}^{H}\left(z^{-1}\right)\mathbf{E}(z)=\mathbf{I} \tag{25}$$

where the H superscript denotes the conjugated transpose, and $\mathbf{I}$ denotes the identity matrix. Paraunitary filter banks also have an attractive property of losslessness, which implies that for every frequency, the total signal power is conserved [16]. From this property [17], any $M\times M$ real-coefficient lossless matrix with $N-1$ degree can be realized using the structure shown in **Figure 9** [18].

If the real-coefficient lossless matrix is denoted by $\mathbf{E}(z)$; then the matrix is said to have a special case of lossless degree of one iff it can be characterized by the relationship [18]:

$$\mathbf{E}(z)=\left[\mathbf{I}-\mathbf{v}\mathbf{v}^{+}+z^{-1}\mathbf{v}\mathbf{v}^{+}\right]\mathbf{R} \tag{26}$$

where $\mathbf{R}$ is an arbitrary $M\times M$ unitary matrix and $\mathbf{v}$ is an $M\times 1$ column vector with unit norm. From Eq. (26), the paraunitary condition for a filter bank is obtained as follows [18]:

$$\left[\mathbf{I}-\mathbf{v}_k\mathbf{v}_k^{+}+\mathbf{v}_k\mathbf{v}_k^{+}z\right]\mathbf{E}_k(z)=\mathbf{E}_{k-1}(z) \tag{27}$$

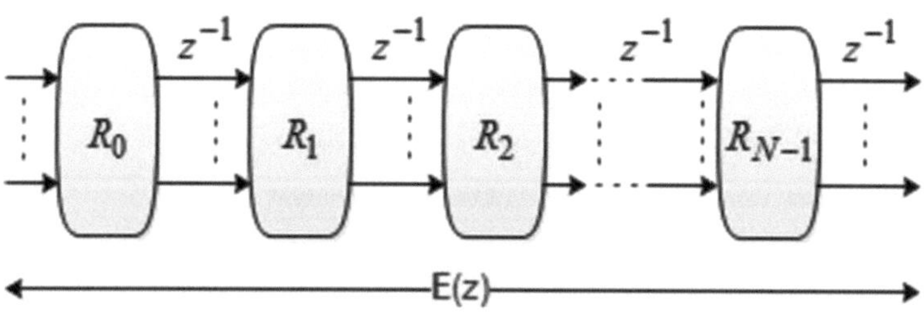

Figure 9.
Cascade implementation of $\mathbf{E}(z)$ *as FIR lossless unitary matrices separated by delays.*

Having looked at the filter bank and its three important properties for the design of a wavelet, we will in the next section examine the application of these properties in the design of a wavelet.

5. Filter bank design of a wavelet transform

The filter bank design of a wavelet transform is usually implemented from the analysis filter bank segment to the synthesis filter bank segment.

5.1 Analysis filter bank in wavelet transform design

Given that the expression for a scaling function $\varphi[n]$ is the series sum of the shifted versions of $\varphi[2n]$, then according to [15, 16], $\varphi[n]$ can be represented as:

$$\varphi[n] = \sum_k h[k]\sqrt{2}\varphi(2n-k), \forall k \in \mathbb{Z} \tag{28}$$

where $h[k]$ denotes the scaling coefficients. If n is transformed such that $n \to 2^\alpha n - \beta$, then the relationship in Eq. (28) becomes [14]:

$$\varphi[2^\alpha n - \beta] = \sum_k h[k]\sqrt{2}\varphi[2(2^\alpha n - \beta) - k] \tag{29}$$

which translates into:

$$\varphi[2^\alpha n - \beta] = \sum_{m=2\beta+k} h[m-2\beta]\sqrt{2}\varphi\left[2^{\alpha+1}n - m\right] \tag{30}$$

when $k = m - 2\beta$.

In a similar consideration to Eq. (28), the wavelet function $\psi[n]$ can be represented as [19]:

$$\psi[n] = \sum_k g[k]\sqrt{2}\varphi(2n-k), \forall k \in \mathbb{Z} \tag{31}$$

where $g[k]$ denotes the wavelet coefficients. Also, if n is transformed such that $n \to 2^\alpha n - \beta$, then the relationship in Eq. (31) becomes [14]:

$$\varphi[2^\alpha n - \beta] = \sum_k g[k]\sqrt{2}\varphi[2(2^\alpha n - \beta) - k] \tag{32}$$

which translates into:

$$\psi[2^\alpha n - \beta] = \sum_{m=2\beta+k} g[m-2\beta]\sqrt{2}\varphi\left[2^{\alpha+1}n - m\right] \tag{33}$$

when $k = m - 2\beta$.

5.2 Synthesis filter bank in wavelet transform design

In the synthesis filter bank, the reconstruction of the original coefficients of a signal can be achieved through the combination of the scaling and wavelet function coefficients at a coarse level of resolution. Given a signal at $\alpha + 1$ scaling space $f[n] \in V_{\alpha+1}$, then according to [16, 17], the reconstruction is derived as follows:

$$f[n] = \frac{1}{\sqrt{M}}\left(\sum_{\beta=-\infty}^{\infty} \lambda_{\alpha+1,\beta}\varphi_{\alpha+1,\beta}[n]\right) = \frac{1}{\sqrt{M}}\left(\sum_{\beta=-\infty}^{\infty} \lambda_{\alpha+1,\beta}\sqrt{2^{\alpha+1}}\psi\left[2^{\alpha+1}n-\beta\right]\right) \tag{34}$$

For the next scale, Eq. (34) becomes:

$$f[n] = \frac{1}{\sqrt{M}}\left(\sum_{\beta} \lambda_{\alpha,\beta}2^{\alpha/2}\varphi[2^{\alpha}n-\beta] + \sum_{\beta} \gamma_{\alpha,\beta}2^{\alpha/2}\psi[2^{\alpha}n-\beta]\right) \tag{35}$$

Substituting Eqs. (28) and (31) into Eq. (35) and after algebraic manipulations yields [14]:

$$\lambda_{\alpha+1,\beta} = \sum_{m} \lambda_{\alpha,\beta}h[\beta-2m] + \sum_{m} \gamma_{\alpha,\beta}g[\beta-2m] \tag{36}$$

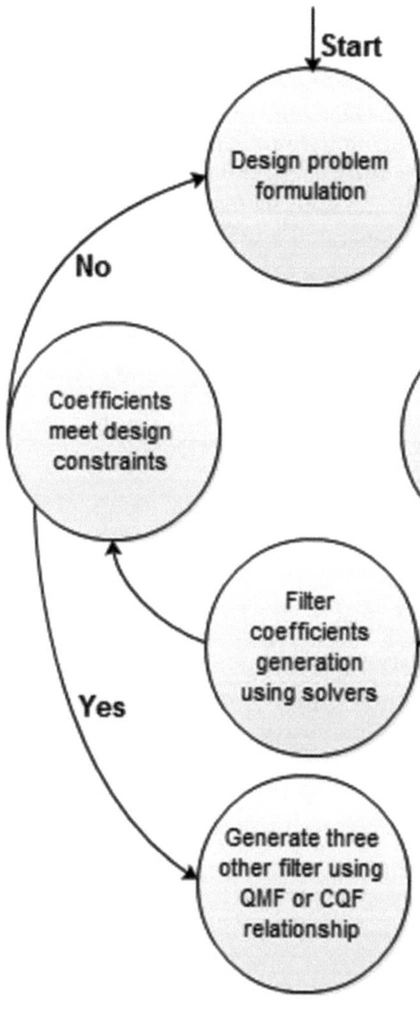

Figure 10.
State chart for wavelet design procedure.

6. Wavelet transform design procedure using filter banks

In the design of a wavelet system using filter banks, it is of utmost importance that the filters which will execute the filter bank system as shown in **Figure 1**, possess the properties discussed in Section 4. Owing to the fact that in a filter bank, all the filters can be derived from an initial filter H_0 as described in Eq. (13), then this initial filter must be designed in such a manner that the relationships in Sections 5.1 and 5.2 are realized. To this end, the following steps as shown in the state diagram in **Figure 10** are necessary.

In the first state in **Figure 10**, the design problem formulation which can be achieved using trigonometric polynomial, takes the following into consideration [14]:

i. Compact support which guarantees that the wavelet is characterized by finite non-zero coefficient.

ii. Paraunitary condition which guarantees the generation of orthonormal wavelets.

iii. Flatness/k-regularity which guarantees the smoothness of the wavelet in both time and frequency domains.

The second state which involves conditioning the problem as a tractable problem involves, if necessary, transforming a non-linear formulation of the problem to a linear formulation, and then optimizing the problem using techniques like convex optimization. The generation of the filter coefficients using solvers in the third state of the machine involves techniques like spectral factorization. Through simulation in the fourth state of the chart, the generated coefficients can be verified whether or not they meet the design constraints. Using the QMF or CQF relationships in Eqs. (13) and (14), the other filters in the filter bank are generated in the fifth state of the chart.

7. Conclusion

In this chapter, we have presented an analysis of the design of wavelets using filter bank technique. The chapter looked at the two major components of a filter bank which the analysis and the synthesis components. The properties of filter banks which are desirable in the design of wavelets were also investigated, alongside the mathematical description of these properties. The chapter also gave a brief mathematical description of the role the analysis and the synthesis filter banks play in the design of wavelets. Finally, the required general procedure for the design of wavelets was presented, showing the necessary steps to take in order to achieve an effective design.

The major contribution of this chapter is the provision of a step by step analysis and procedure for the design of filter banks in a precise and concise manner.

Author details

Peter Yusuf Dibal[1*], Elizabeth Onwuka[1], James Agajo[2] and Caroline Alenoghena[1]

1 Telecommunication Engineering Department, Federal University of Technology Minna, Minna, Nigeria

2 Computer Engineering Department, Federal University of Technology Minna, Minna, Nigeria

*Address all correspondence to: yoksa77@gmail

References

[1] Mertins A. Signal Analysis: Wavelets, Filter Banks, Time-Frequency Transforms and Applications. Sussex: Wiley; 1999

[2] Mon AT, Aye SM, Tun HM, Naing ZM, Moe WK. Analysis of multichannel filter banks-based tree-structured design for communication system. Software Engineering. 2018;**6**(2):37-46

[3] Mallat S. A Wavelet Tour of Signal Processing: The Sparse Way. 3rd ed. Burlington: Academic Press; 2009

[4] Tay DB, Ortega A, Anis A. Cascade and lifting structures in the spectral domain for bipartite graph filter banks. In: APSIPA Annual Summit and Conference. Cambridge University Press; 2018. pp. 1141-1147

[5] Chu E. Discrete and Continuous Fourier Transforms: Analysis, Applications and Fast Algorithms. Boca Raton: CRC Press; 2008

[6] Abbaszadeh MH, Khalaj BH, Haghbin A. Optimum low complexity filter bank for generalized orthogonal frequency division multiplexing. EURASIP Journal on Wireless Communications and Networking. 2018;**2018**(12):1-14

[7] Vantita S, Rajesh M. Rational sampling rate converter design analysis using symmetric technique. International Journal of Computer Trends and Technology. 2015;**27**(2): 116-120

[8] Devis T, Manuel M. Multirate and filterbank approaches in digital hearing aid design: A review. In: IOP Materials Science and Engineering Conference. IOP Publishing; 2018. pp. 1-8

[9] Agrawal SK, Sahu OP. Two channel quadrature mirror filter bank: An overview. Hindawi—ISRN Signal Processing. 2013;**2013**:1-10

[10] Giron-Sierra JM. Digital Signal Processing with MATLAB Examples. Singapore: Springer Singapore; 2017

[11] Vetterli M, Herley C. Wavelets and filter banks: Theory and design. IEEE Transactions on Signal Processing. 1992; **40**(9):2207-2231

[12] Vaidyanathan PP. Multirate Systems and Filter Banks. New Jersey: Prentice-Hall; 1993

[13] Gao XQ, Nguyen TQ, Strang G. A study of two-channel complex-valued filterbanks and wavelets with orthogonality and symmetry properties. IEEE Transactions on Signal Processing. 2002;**50**(4):824-833

[14] Nikookar H. Wavelet Radio: Adaptive and Reconfigurable Wireless Systems Based on Wavelets. Cambridge: Cambridge University Press; 2013

[15] Gao X, Nguyen TQ, Strang G. On factorization of M-channel paraunitary filterbanks. IEEE Transactions on Signal Processing. 2001;**49**(7): 1433-1466

[16] Redif S, McWhirter JG, Weiss S. Design of FIR paraunitary filter banks for subband coding using a polynomial eigenvalue decomposition. IEEE Transactions on Signal Processing. 2011; **59**(11):5253-5264

[17] Vaidyanathan PP, Dongata Z, Nguyen TQ. More results on the perfect reconstruction problem in M-band parallel QMF bank. In: IEEE International Symposium on Circuits and Systems. IEEE; 1987. pp. 847-850

[18] Vaidyanathan PP, Nguyen TQ, Doganata Z, Saramaki T. Improved

technique for design of perfect reconstruction FIR QMF banks with lossless polyphase matrices. IEEE Transactions on Acoustics, Speech, and Signal Processing. 1989;**37**(7):1042-1056

[19] Burrus CS, Gopinath RA, Guo H. Introduction to Wavelets and Wavelet Transform: A Primer. Upper Saddle River: Prentice Hall; 1998

Chapter 3

Wavelets for Differential Equations and Numerical Operator Calculus

Riccardo Bernardini

Abstract

Differential equations are commonplace in engineering, and lots of research have been carried out in developing methods, both efficient and precise, for their numerical solution. Nowadays the numerical practitioner can rely on a wide range of tools for solving differential equations: finite difference methods, finite element methods, meshless, and so on. Wavelets, since their appearance in the early 1990s, have attracted attention for their multiresolution nature that allows them to act as a "mathematical zoom," a characteristic that promises to describe efficiently the functions involved in the differential equation, especially in the presence of singularities. The objective of this chapter is to introduce the main concepts of wavelets and differential equation, allowing the reader to apply wavelets to the solution of differential equations and in numerical operator calculus.

Keywords: wavelets, differential equations, numerical analysis, finite element method, meshless, multiresolution analysis

1. Introduction

Partial differential equations (PDEs) are used commonplace in science and in engineering to model the behavior of physical systems. Because of their importance, many numerical techniques for their solutions have been developed: finite difference methods (FDMs), finite element methods (FEMs), spectral methods, Ritz/Galerkin approach, meshless approaches, and so on. The main characteristic of PDEs is that the "unknown" is a function, that is, an object with an infinite number of degrees of freedom. Because of this, it is usually impossible (even in principle) to get an exact solution by numerical means. The objective of every technique for PDE solving is to get a good approximation of the solution with limited computational resources (CPU time, memory, etc.).

The first step of every PDE solution algorithm is to *discretize* the PDE, that is, to approximate it with a finite-dimensional problem that can be solved by numerical means. A popular discretization technique is to discretize the space where the solution is searched by restricting the problem to a finite-dimensional vector space, that is, by writing the solution as the linear combination of several base functions. If the original PDE was linear, discretization will map it to a linear problem (typically a linear system or an eigenvalue problem). Even techniques such as FDM (that discretizes the domain) can be often reformulated as suitable discretization of the function space.

Intuitively, better approximations of the solution require *finer* discretization (the exact meaning of *finer* depends on the specific approach), especially if the solution has some regions of large variability. Since finer discretization implies larger problems (and, therefore, higher computational efforts), it is of interest to be able to change locally the discretization resolution to the solution variability, possibly in an adaptive way.

This need for different resolutions in different regions is the idea that links PDE with *multiresolution analysis*. The birth of multiresolution analysis goes back to 1990 with the works of Mallat [1] and Meyer [2]. Since then there have been a large number of papers ranging from very theoretical ones to application [3]. In a multiresolution analysis, a space of signals (most commonly $L^2(\mathbb{R})$, but not only) is represented as a nesting of spaces with different levels of "resolution." This allows to write a signal as the sum of a "low-resolution" version plus some higher-resolution "details."

Because of the ability of changing the resolution used to observe the signal (by adding or removing details), the multiresolution analysis is sometimes described as a *mathematical zoom*. This fact inspired many applications, including numerical solution of PDEs where they sound promising, especially for those problems that contains localized phenomena (e.g., shockwaves) or intra-scale interaction (e.g., turbulence).

The objective of this chapter is to introduce the reader to the application of wavelets to PDE solutions. This chapter can be ideally divided in three parts: in the first part, we recall briefly the main concepts about PDE and the main algorithms for solving PDE; successively we do a brief recall of multiresolution analysis and wavelets including also multiwavelets and second-generation wavelets that find often application in PDE solutions; and finally, we will illustrate few techniques that can be found in the literature.

1.1 Notation

$\Omega \subseteq \mathbb{R}^d$ is the domain where the functions of interest are defined. The boundary of Ω is partitioned as follows: $\partial\Omega = \Omega_D \cup \Omega_N$, $\Omega_D \cap \Omega_N = \emptyset$, where Ω_D or Ω_N can be empty.

$\mathcal{H}(\Omega)$ will denote a space of functions $u : \Omega \to \mathbb{R}$ defined on $\Omega \subseteq \mathbb{R}^d$.

$\mathbb{R}_{\geq 0}$ is the set of nonnegative reals.

2. Generalities on PDE

Most of the physical problems modeled by PDEs fall in one of the following three large classes: equilibrium, propagation, or eigenvalue problems.

- In an *equilibrium problem*, we are interested in finding a function $u \in \mathcal{H}(\Omega)$ such that

$$\mathcal{D}u = f \quad \text{in } \Omega \tag{1}$$

$$u = u_D \quad \text{on } \Omega_D \tag{2}$$

$$\frac{\partial u}{\partial \mathbf{n}} = u_N \quad \text{on } \Omega_N \tag{3}$$

In (1)–(3) $\mathcal{D}$ is an operator that includes derivatives and $f \in \mathcal{H}(\Omega)$ is known. Boundary conditions are typically given as constraint about u or its derivatives on regions of $\partial\Omega$. Typical examples of physical systems giving rise to this type of problem are systems in steady state (e.g., temperature distribution, potential distribution, steady flows, and so on). Typically equilibrium problems are *elliptic*, that is, $\mathcal{D}$ is an *elliptic operator* (a generalization of the Laplacian).

- In a *propagation problem*, we are interested in modeling the time evolution of a physical system. The PDE can still be written as in (1)–(3), but the domain can typically be written as $\Omega = \mathbb{R}_{\geq 0} \times W$, where $W \subseteq \mathbb{R}^{d-1}$ and the first coordinate represents time. Boundary conditions for $t = 0$ are known as *initial conditions*. Example physical problems are heat or wave propagation. Propagation problems are typically hyperbolic or parabolic.

- Finally, in *eigenvalue problems* we are interested in finding u and λ that satisfy

$$\mathcal{D}u = \lambda u \tag{4}$$

A wide class of eigenvalue problems is represented by *Sturm-Liouville problems* that can be written as

$$[p \cdot y']' + q \cdot y = -\lambda w \cdot y \tag{5}$$

where the apostrophe denotes derivation, the unknowns are λ, and function is $y \in \mathcal{H}([a, b])$, while p, q, and w, all belonging to $\mathcal{H}([a, b])$, are known. Sturm-Liouville problems include Bessel differential equations (obtained by writing Laplace, Helmholtz, or Schrodinger equation in cylindrical coordinates) and Lagrange differential equation (obtained working in spherical coordinates).

2.1 Solution of differential equations

The field of numerical solution of differential equations is very wide, and many techniques have been developed. Nevertheless, a categorization in few large classes is possible. An important step in every solution algorithm is mapping the differential equation into a discrete version with only a finite number of degrees of freedom. A first distinction can be done between techniques that achieve this objective by discretizing the domain Ω or the function space $\mathcal{H}(\Omega)$.

2.1.1 Domain discretization

The most known technique based on a domain discretization is the FDM where the unknown function is sampled in a finite number of points $p_1, p_2, \ldots, p_N \in \Omega$ and the derivatives are approximated with finite differences. By writing the differential equation for every p, with the derivative approximated as finite differences, one obtains a system of N equations in N unknowns that can be solved with known techniques. If the original PDE was linear, the discretized system will be linear too.

FDM is maybe the simplest approach and the most intuitive, and it can work quite well for simple problems and geometries. Moreover, in the linear case, since any approximation of a derivative in p will consider only few points around p, the matrix of the discretized linear system will be very sparse, allowing for a reduction in the computational effort. The application of FDM techniques becomes difficult, albeit possible, in the case of complex problems.

2.1.2 Function space discretization

Another class of techniques discretizes the function space $\mathcal{H}(\Omega)$ by approximating it with an n-dimensional space H_n, that is, unknown function u is approximated as

$$u \approx \sum_{i=1}^{n} \alpha_i b_i, \qquad \alpha_i \in \mathbb{R} \tag{6}$$

where $\{b_i\}_{i=1}^{n}$ is a basis of H_n.

By exploiting approximation (6), one can transform PDE (1)–(3) into a finite-dimensional problem. The different solution techniques differ in how (6) is used and in the way of choosing space H_n and its basis $\{b_i\}_{i=1}^{n}$.

One possibility is to choose functions b_i that are infinitely differentiable and nonvanishing on the whole Ω. This gives rise to so-called *spectral* methods. Typical choices for basis functions can be complex exponential/sinusoidal functions (if the solution is expected to be periodic), Chebyshev polynomials (for separable domains, e.g., d-dimensional cubes), and spherical harmonics (for systems with spherical symmetry). Spectral methods can work very well if the solution is expected to be smooth; they can even converge exponentially fast. However, their spatial localization is not good, and if the functions involved are not smooth (e.g., they are discontinuous), they lose most of their interest.

Another approach, very popular, is FEM that chooses functions b_i by first partitioning the domain Ω into a set of *elements* (triangles and their multidimensional counterpart are a popular choice) and assigning to every element a suitable finite-dimensional vector space. The final approximation of u is constructed in a piecewise fashion by gluing, so to say, the approximations of u over every single element.

In a typical implementation of FEM, all the elements are affine images of a single reference element. This simplifies the implementation since it suffices to choose only the vector space of the reference element T_0. Another popular choice is to choose the space associated to the elements as spaces of polynomials. The basis is selected by choosing a set of control points in $q_1, q_2, \ldots \in T_0$ and choosing as basis vectors b_i the polynomials that satisfy the *interpolation property*

$$b_i\left(q_j\right) = \delta_{i,j} = \begin{pmatrix} 1 & \text{if } i = j \\ 0 & \text{if } i \neq j \end{pmatrix} \tag{7}$$

Remark 2.1 (generalized collocation method).

A generalization of this idea is to choose a set of functionals σ_j mapping functions defined over T_0 to $\mathbb{R}$ and requiring

$$\sigma_j(b_i) = \delta_{i,j} \tag{8}$$

Eq. (8) gives back (7) if σ_j is defined as the functional that corresponds to evaluating the argument of the functional in q_j. Eq. (8) is, however, more general than (7) since it can be used, for example, to control the flow through a face of the element.

An issue with FEM is that creating the grid of elements can be expensive. This is especially true in those problems where the geometry is not fixed but needs to be updated. An example of this type of system is free-surface fluid flows, where the interface between air and fluid changes with time, requiring a continuous update of the mesh. In order to solve this problem, *meshless* methods have been developed [4]. A typical meshless approach is to approximate u with a discrete convolution with kernel as a chosen function φ, that is,

$$u(x) = C_\rho \sum_I \alpha_I \phi \left(\frac{x - x_I}{\rho} \right) \tag{9}$$

where X_I are a set of points of Ω and ρ is a scale factor that allows to change the "resolution" of kernel function ϕ. Coefficient C_ρ can be used to keep the energy constant as ρ is changed.

2.1.3 Exploiting the discretization

After expressing u as linear combination of b_i, we are left with the problem of determining the coefficients of the linear combination. Several approaches are possible; the easiest way to briefly present them is by rewriting the differential equation as

$$\mathcal{R}u := \mathcal{D}u - f = 0 \tag{10}$$

where operator $\mathcal{R} : \mathcal{H}(\Omega) \to \mathcal{H}(\Omega)$ is called the *residual*.

If we restrict u to be a linear combination of b_i, most probably we will not be able to make residual (10) exactly zero; therefore, we will aim to make it as small as possible. Since the result of the residual operator is a function, there are many possible approaches in minimizing it.

With the *collocation* approach, we choose a number of points of the domain $p_1, p_2, \ldots, p_n \in \Omega$ and ask that the residual is zero on the chosen points, that is,

$$0 = [\mathcal{R}u]\left(p_j\right) = [\mathcal{D}u]\left(p_j\right) - f\left(p_j\right) \qquad j = 1, \ldots, n \tag{11}$$

Eq. (11) represents a system of n equations having as unknown the coefficients α_i, $i = 1, \ldots, n$. For example, if $\mathcal{D}$ is linear, (11) becomes

$$f\left(p_j\right) = \left[\mathcal{D} \sum_{i=1}^{n} \alpha_i b_i\right]\left(p_j\right) = \sum_{i=1}^{n} \alpha_i [\mathcal{D}b_i]\left(p_j\right) = \sum_{i=1}^{n} \alpha_i A_{j,i} \qquad j = 1, \ldots, n \tag{12}$$

where, clearly, $A_{j,i} = [\mathcal{D}b_i]\left(p_j\right)$. Note that (12) is a linear system in unknowns α_i.

Remark 2.2.

With reference to Remark 2.1, one can generalize the collocation method by using a set of linear functionals $\sigma_j : \mathcal{H}(\Omega) \to \mathbb{R}$. In this case one can obtain a generalized version of (12), namely,

$$\sigma_j f = \sum_{i=1}^{n} \alpha_i \underbrace{\sigma_j(\mathcal{D}b_i)}_{A_{j,i}} \qquad j = 1, \ldots, n \tag{13}$$

Another approach is to solve $\mathcal{R}u = 0$ in a least square sense, that is, to search for coefficients α_i that minimize

$$\|\mathcal{R}u\|^2 = \langle \mathcal{R}u, \mathcal{R}u \rangle \tag{14}$$

Standard algebra allows to show that (14) is minimized when $\mathcal{R}u$ is orthogonal to $\partial \mathcal{R}u / \partial \alpha_i$ for every i, that is,

$$\left\langle \mathcal{R}u, \frac{\partial \mathcal{R}u}{\partial \alpha_i} \right\rangle = 0 \qquad i = 1, \ldots, n \tag{15}$$

If $\mathcal{D}$ is linear,

$$\frac{\partial \mathcal{R}u}{\partial \alpha_j} = \frac{\partial}{\partial \alpha_j} \left[\mathcal{D} \sum_{i=1}^{n} \alpha_i b_i - f \right] = \mathcal{D} b_j \tag{16}$$

and we get

$$\begin{aligned} 0 &= \left\langle \mathcal{R}u, \frac{\partial \mathcal{R}u}{\partial \alpha_j} \right\rangle \\ &= \left\langle \mathcal{D} \sum_{i=1}^{n} \alpha_i b_i - f, \mathcal{D} b_j \right\rangle \\ &= \sum_{i=1}^{n} \alpha_i \langle \mathcal{D} b_i, \mathcal{D} b_j \rangle - \langle f, \mathcal{D} b_j \rangle \end{aligned} \tag{17}$$

which is still a linear system.

The *Galerkin method* is inspired on the idea that in a least square approximation, the error is orthogonal to the space where the approximating function lives. We would like to approximate the solution of the PDE with a vector of H_n; however, we do not know the solution, so we ask for the *residual* to be orthogonal to H_n, that is,

$$\langle \mathcal{R}u, v \rangle = 0 \qquad \forall v \in H_n \tag{18}$$

Eq. (18) is equivalent to

$$\langle \mathcal{D}u, v \rangle = \langle f, v \rangle \qquad \forall v \in H_n \tag{19}$$

which can be interpreted as the original differential equation $\mathcal{D}u = f$ in *weak form*. Form (19) is often exploited by integrating by parts the left-hand side scalar product, moving one differentiation from the unknown function u to the test function v. This is often useful when a piecewise linear approximation is employed and $\mathcal{D}$ contains second-order differential operators (that cannot be applied on piecewise linear functions). Eq. (19) is verified for all $v \in H_n$ if and only if it is verified for every vector in a basis of H_n, that is, (19) is equivalent to

$$\langle \mathcal{D}u, b_j \rangle = \langle f, b_j \rangle \qquad j = 1, \ldots, n \tag{20}$$

If $\mathcal{D}$ is linear, from (20) one can easily derive the linear system in α_i

$$\langle f, b_j \rangle = \sum_{i=1}^{n} \alpha_i \langle \mathcal{D} b_i, b_j \rangle \qquad j = 1, \ldots, n \tag{21}$$

Finally, it is worth citing the *method of weighted residuals* that can be seen as a generalization of the Galerkin PDE method. The idea is that instead of asking the residual being orthogonal to the space H_n used to approximate u, we ask the residual to be orthogonal to a different n-dimensional space $K_n = \text{span}\{\beta_1, \ldots, \beta_n\}$ where $\{\beta_i\}_{i=1}^{n}$ is clearly a basis of K_n. One obtains

$$\langle f, \beta_j \rangle = \sum_{i=1}^{n} \alpha_i \langle \mathcal{D} b_i, \beta_j \rangle \qquad j = 1, \ldots, n \tag{22}$$

Remark 2.3.

It is worth observing that from the weighted residual method, Galerkin and least square methods can be derived by a suitable choice of β_i; even collocation method can be derived if we allow β_i to be a delta function (so that the scalar product needs to be interpreted as a distribution pairing). Moreover, for every v since map $x \mapsto \langle x, v \rangle$ is a functional, it is easy to recognize that every method can be considered like a *generalized collocation method*, as described in Remark 2.1.

3. Wavelets

The idea of multiresolution analysis is to approximate vectors of $L^2(\mathbb{R})$ with variable degrees of resolution. This is achieved through a *multiresolution analysis scheme* defined by means of some axioms. The first axiom is the existence of a sequence $\{V_n\}_{n \in \mathbb{Z}}$ of subspaces of $L^2(\mathbb{R})$ nested one inside the other, that is,

$$\cdots \subset V_{-2} \subset V_{-1} \subset V_0 \subset V_1 \subset V_2 \subset \cdots \tag{23}$$

The idea is that if one approximates (in a least square sense) a function f with vectors belonging to V_n, the approximation error gets smaller as n increases since every vector of V_n also belongs to V_{n+1}. Note, however, that (23) does not grant that we will be able to approximate f with an error as small as desired; in order to grant this, we need another axiom

$$\overline{\bigcup_{n \in \mathbb{Z}} V_n} = L^2(\mathbb{R}) \tag{24}$$

where the overline denotes set closure (in the topology induced by the norm on $L^2(\mathbb{R})$). Axiom (24) requires that every vector of $L^2(\mathbb{R})$ is in the closure of the union in the left hand; this means that given any $\epsilon > 0$ and $f \in L^2(\mathbb{R})$, it is possible to find an element of the union whose distance from f is less than ϵ. In other words, (24) means that whatever $f \in L^2(\mathbb{R})$ and whatever the chosen maximum approximation error allowed ϵ, one can find a space V_n that approximates f with the required precision.

An axiom dual to (24) is

$$\bigcap_{n \in \mathbb{Z}} V_n = \{0\} \tag{25}$$

that requires that there is only one "lowest resolution vector," that is, the null vector.

Remark 3.1.

In order to see that axiom (24) is not obvious, it is more convenient to work with Hilbert space $L^2([0,1])$. Recall that functions $x \mapsto \cos(2\pi n x)$, $x \mapsto \sin(2\pi n x)$, and $n \in \mathbb{N}$ and the constant 1 are an orthogonal basis of $L^2([0,1])$.

Define $S_0 = \{\sin(2\pi(2k)t), k \in \mathbb{N}\}$ as the set of all the even-numbered sines, and define V_0 as the space generated by S_0, that is,

$$V_0 := \operatorname{span} S_0 \tag{26}$$

Now define spaces V_n, $n < 0$ by removing one vector at time from the basis of V_0, and define spaces V_n, $n < 0$ by adding one odd harmonic at time. More precisely, define

$$V_n = \operatorname{span} S_n \tag{27}$$

where

$$S_n = \begin{cases} S_{n+1} \{ \sin (2\pi(-2n)t)\} & \text{if n} < 0 \\ S_{n-1} \cup \{ \sin (2\pi(2n-1)t)\} & \text{if n} > 0 \end{cases} \tag{28}$$

It is clear that the sequence of spaces defined in this way satisfies axiom (23), but not (24), since, for example, function $\cos(2\pi t)$ is orthogonal to every V_n. Note that this construction can be repeated for any Hilbert space using an orthonormal basis of the space instead of sines and cosines. Another axiom makes more precise the idea of "increasing resolution" by asking that vectors in V_n vary twice as faster than the vectors in V_{n-1}. In order to make this more precise, define operator $\mathcal{S} : L^2(\mathbb{R}) \to L^2(\mathbb{R})$ as the rescaling operator $[\mathcal{S}f](x) = \sqrt{2}f(2x)$. Note that because of the multiplication by $\sqrt{2}$, $\mathcal{S}$ is unitary, that is, $\|\mathcal{S}f\| = \|f\|$. The new axiom is

$$f \in V_n \Leftrightarrow \mathcal{S}f \in V_{n+1} \tag{29}$$

It follows that $V_n = \mathcal{S}^n V_0$. With this position one can interpret (25) by saying that the slowest function is the constant (and the only constant in $L^2(\mathbb{R})$ is the zero).

The last axiom puts a constraint on the structure of V_0 by asking that is generated by a function ϕ and its translations. In order to make this more precise, define the operator $\tau^t : L^2(\mathbb{R}) \to L^2(\mathbb{R})$ associated to a translation of t as $[\tau^t f](x) = f(x-t)$. Note that also τ^t is unitary and that the exponential notation is convenient since $\tau^a \tau^b = \tau^{a+b}$. Observe also the commutation relation $\mathcal{S}\tau^t = \tau^{t/2}\mathcal{S}$. The last axiom can be written as

$$\exists \phi \in L^2(\mathbb{R}) : V_0 = \operatorname{span}\{\tau^i \phi, \quad i \in \mathbb{Z}\} \tag{30}$$

Often as part of axioms, it is required that ϕ is orthogonal to its translations, that is,

$$\langle \tau^i \phi, \tau^j \phi \rangle = \delta_{i,j} \tag{31}$$

However, it is not necessary to include (31) explicitly in the axioms since, given a ϕ that satisfies (30), it is possible to orthonormalize it, so that it satisfies (31), with a well-known "Fourier trick" [3]. Therefore, we will suppose (31) satisfied.

It is worth to summarize here the axioms

$$\cdots \subset V_{-2} \subset V_{-1} \subset V_0 \subset V_1 \subset V_2 \subset \cdots \tag{32}$$

$$\overline{\bigcup_{n \in \mathbb{Z}} V_n} = L^2(\mathbb{R}) \tag{33}$$

$$\bigcap_{n \in \mathbb{Z}} V_n = \{0\} \tag{34}$$

$$V_{n+1} = \mathcal{S} V_n \tag{35}$$

$$V_0 = \operatorname{span}\{\tau^i \phi, \quad i \in \mathbb{Z}\} \quad \exists \phi \in L^2(\mathbb{R}) \tag{36}$$

The axioms above allow us to determine a property of ϕ. Note that since $V_1 \supset V_0$, $\phi \in V_1$. Note also that set $\{\mathcal{S}\tau^i \phi = \tau^{i/2} \mathcal{S}\phi i \in \mathbb{Z}\}$ is an orthonormal basis of V_1. It follows that one can write ϕ as linear combination of $\mathcal{S}\phi$ and its half-integer translations, that is,

$$\phi = \sum_{i \in \mathbb{Z}} g_i \; \tau^{i/2} \mathcal{S} \phi \tag{37}$$

for some sequence $g_i : \mathbb{Z} \to \mathbb{R}$. Eq. (37) is known as *two-scale equation* and it is central to wavelet theory. Function ϕ is known as *scaling function*.

Remark 3.2.

Note that from the orthonormality of $\{\tau^{i/2} \mathcal{S} \phi i \in \mathbb{Z}\}$ follows

$$\langle g_i, \tau^{2k} g_i \rangle = \langle \varphi, \tau^{-k} \varphi \rangle = \delta_{k,0} \tag{38}$$

where the left-hand side scalar product is the usual scalar product in $\ell^2(\mathbb{Z})$.

Remark 3.3.

Note that starting from a ϕ that satisfies a two-scale equation like (37), it is possible to recover a full multiresolution analysis. Indeed, one defines V_0 according to (36) and V_n by repeated applications of (35). Two-scale Eq. (37) grants that the nesting axiom (32) is satisfied.

Note also that (37) shows that ϕ is the fixed point of operator

$$\mathcal{O} := \sum_{i \in \mathbb{Z}} g_i \; \tau^{i/2} \mathcal{S} \tag{39}$$

This suggests that maybe one could start from a sequence g_i and apply repeatedly $\mathcal{O}$ to a vector of $L^2(\mathbb{R})$ in order to obtain ϕ. This is indeed possible, but the theoretical details are out of scope here; see [5].

Since $V_{n+1} \supset V_n$ one can consider the orthogonal complement of V_n in V_{n+1}; call it W_n, that is,

$$V_{n+1} = V_n \oplus W_n \tag{40}$$

It is possible to find, starting from the two-scale Eq. (37), a function ψ such that $\{\tau^i \psi\}_{i \in \mathbb{Z}}$ is an orthonormal basis of W_0. This implies that it must be for all

$$\psi = \sum_{i \in \mathbb{Z}} h_i \; \tau^{i/2} \mathcal{S} \phi \quad \psi \in V_1 \tag{41}$$

$$\langle \psi, \tau^i \psi \rangle = \delta_i \qquad \text{orthonormal basis} \tag{42}$$

$$\langle \tau^j \psi, \tau^i \phi \rangle = 0 \qquad \forall i, j \in \mathbb{Z} \qquad W_0 \text{ orthogonal to } V_0 \tag{43}$$

By using (41) and the orthonormality of $\tau^{i/2} \mathcal{S} \phi$, it is possible to rewrite (42) and (43) as conditions on h_i, namely,

$$\langle \tau^{2k} h_i, h_i \rangle = \delta_{k,0} \tag{44}$$

$$\langle \tau^{2k} h_i, g_i \rangle = 0 \tag{45}$$

It is easy to verify that, given g_i, a possible h_i that satisfies (44) and (45) is

$$h_i = (-1)^i g_{-n+1} \tag{46}$$

This shows that sequences h_i and g_i are the impulse responses of a two-channel orthogonal filter bank. Moreover, if g_i and ϕ are known, one can obtain ψ by choosing h_i according to (46) and computing ψ according to (41). Function ψ is known as *wavelet*, and it generates the whole $L^2(\mathbb{R})$ with its translations and dilations.

This has also another interesting consequence. Suppose $f \in L^2(\mathbb{R})$ and that

$$\gamma_i^{(1)} = \left\langle \tau^{i/2} S\phi, f \right\rangle = \left\langle S\tau^i \phi, f \right\rangle \tag{47}$$

are the coefficients of its projection on V_1. Suppose we need the coefficients $\gamma_k^{(0)} = \langle \tau^k \phi, f \rangle$ of the projection on V_0. It is possible to exploit the two-scale equation

$$\begin{aligned} \gamma_k^{(0)} = \langle \tau^k \phi, f \rangle &= \left\langle \tau^k \sum_{i \in \mathbb{Z}} g_i S\tau^i \phi, f \right\rangle \\ &= \sum_{i \in \mathbb{Z}} g_i \langle S\tau^{i+2k} \phi, f \rangle \\ &= \sum_{i \in \mathbb{Z}} g_i \gamma_{i+2k}^{(1)} = \left[g_- * \gamma_i^{(1)} \right]_{2k} \end{aligned} \tag{48}$$

where g_- is the time-reversed version of g_i. Eq. (48) shows that it is possible to go to the space at lower resolution by means of a filtering by g_- and a decimation by a factor of two. Similarly, by calling $\eta_k^{(0)} = \langle \tau^k \psi, f \rangle$ the coefficients relative to the projection of f on W_0, one can obtain

$$\eta_k^{(0)} = \sum_{i \in \mathbb{Z}} h_i \gamma_{i+2k}^{(1)} = \left[h_- * \gamma_i^{(1)} \right]_{2k} \tag{49}$$

Figure 1a shows this idea: the sequence of high-resolution coefficients are processed with a two-channel orthogonal filter bank, and the coefficients relative to the lower resolution space V_n exit from one branch, and the coefficients relative to the “missing details” space W_n exit from the other. The idea can be iterated several times; see **Figure 1b**. This is the basis of the well-known fast algorithm to compute wavelet coefficients and also the origin of the minor, and very common, misnomer in calling **Figure 1b** a “discrete-time wavelet transform.”

An interesting characteristic of wavelets is that they can be used to detect the local regularity of a function. This is similar to what happened with Fourier transform where a function that is discontinuous has a Fourier transform that decays as $1/\omega$; if the function is continuous but not derivable, its Fourier transform decays as $1/\omega^2$ and so on. With the wavelet transform happens something similar, with the scale playing the role of frequency. The interesting difference is that while a Fourier transform that decays as $1/\omega$ tells us that there is at least one discontinuity, but not where, with the wavelet transform the slow decay with the scale is localized around the discontinuity. The precise claim of this property requires the introduction of the concept of Lipschitz regularity and would take us too far; see [6]. This suggests that when approximating the unknown function in a PDE, we can keep high-resolution coefficients only in the neighborhood of singularities, saving on computational effort.

We will say that wavelet ψ has ℓ vanishing moments if

$$\int_{\mathbf{R}} x^k \psi(x) dx = 0 \qquad k = 0, 1, \ldots, \ell - 1 \tag{50}$$

An interesting property of compactly supported wavelets with ℓ vanishing moments is that the *corresponding scaling function* (not the wavelet itself) can reproduce polynomials of degree at most $\ell - 1$ in the sense that if $P(x)$ is a polynomial with degree less than ℓ, there exist coefficients c_i such that

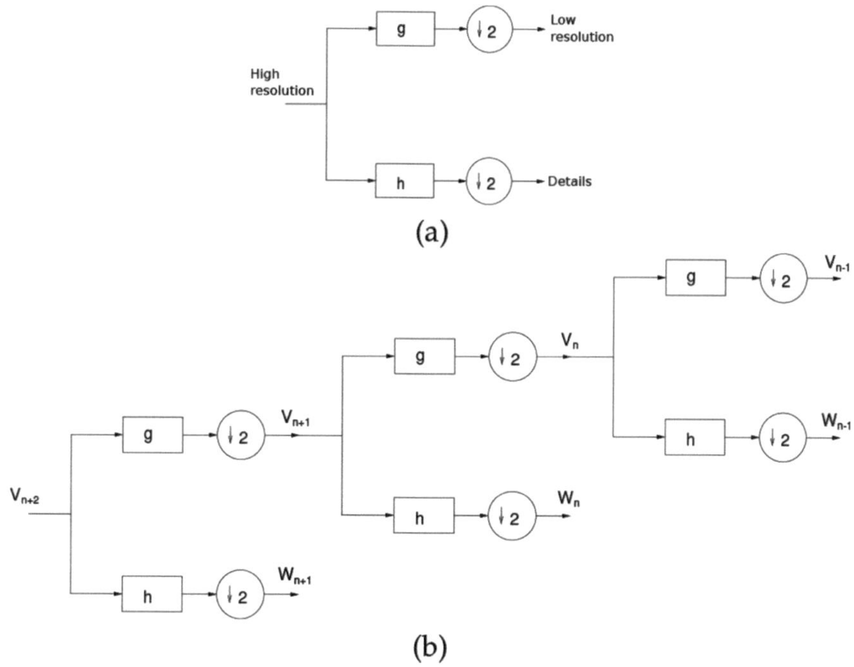

Figure 1.
(a) Splitting coefficient sequence into a low-resolution and a high-resolution one using a two-channel filter bank. (b) Iteration of structure (a) makes a fast algorithm for computing the wavelet coefficients.

$$P(x) = \left[\sum_{i \in \mathbb{Z}} c_i \tau^i \phi\right](x) \tag{51}$$

In other words, V_0 contains all the polynomials of degree less than ℓ.
Example 3.1. (Haar wavelet).
The simplest example of wavelet is the Haar wavelet whose scaling function is

$$\phi_H(x) = \begin{pmatrix} 1 & \text{if} \, x \in [0,1] \\ 0 & \text{else} \end{pmatrix} \tag{52}$$

It is immediate to verify that ϕ_H satisfies a two-scale equation

$$\phi_H = \frac{1}{\sqrt{2}}\left(S\phi_H + \tau^{1/2} S\phi_H\right) \tag{53}$$

with coefficients $g_0 = g_1 = 1/\sqrt{2}$. Note that trivially $\langle \tau^{2k}\phi_H, \phi_H \rangle = \delta_k$. In order to create the corresponding wavelet, use prescription $h_i = (-1)^i g_{-i+1}$ to get

$$\phi_H = \frac{1}{\sqrt{2}}\left(S\phi_H - \tau^{1/2} S\phi_H\right) \tag{54}$$

Note that the Haar wavelet is compactly supported, but it is discontinuous. This makes it not well suited to approximate smooth functions.
Example 3.2 (Sinc wavelet).
An example in some sense opposite to the Haar wavelet is the Sinc wavelet. In this case V_0 is the space of "low-pass" functions, that is, functions whose Fourier transform is zero outside interval $[-\pi, \pi]$. As well known, V_0 is generated by the Sinc function

$$\operatorname{sinc}(x) = \frac{\sin(\pi x)}{\pi x} \tag{55}$$

and its translations, that is,

$$V_0 = \operatorname{span}\{\tau^k \operatorname{sinc}\}_{k\in\mathbb{Z}} \tag{56}$$

This suggests to use $\phi_S = \operatorname{sinc}$ as scaling function. The fact that a two-scale equation is satisfied is easily checked in frequency since V_1 is the space of functions whose Fourier transform is zero outside $[-2\pi, 2\pi]$; therefore, every function of V_0 is contained in V_1, as desired.

The corresponding wavelet ψ_S is easily characterized in frequency as the function whose Fourier transform is

$$\Psi_S(\omega) = \begin{cases} 1 & \text{if } \pi < |\omega| < 2\pi \\ 0 & \text{otherwise} \end{cases} \tag{57}$$

It is easy to verify that $\psi_S \in V_1$, $\psi_S \perp V_1$, and $\langle \tau^{2k}\psi_S, \psi_S\rangle = \delta_k$.

As said above, the Sinc example is somehow the opposite of Haar wavelet: it is arbitrarily differentiable, but it has infinite support; actually, it decays very slowly (as $O(1/x)$), and this introduces several practical issues. Moreover, sequences g_i and h_i are of infinite length, and they decay slowly too (they do not even have a z-transform), making it difficult to implement it.

Example 3.3 (spline wavelet).

An example intermediate between Haar and Sinc wavelet is represented by spline spaces of degree d. In this case $V_0^{(d)}$ is defined as the space of piecewise polynomial functions that are d times differentiable (with continuous derivative), with the "breaking points" on the integer, more precisely

$$V_0^{(d)} = \left\{ f \in L^2(\mathbb{R}) \cap C_{d-1}, f\big|_{[k,k+1]} = \text{polynomial degree } d \, \forall k \in \mathbb{Z} \right\} \tag{58}$$

It is easy to see that $V_1^{(d)} = \mathcal{S}V_0^{(d)}$ is a similar space of piecewise polynomial functions but with the breaking points in half integers. It follows that every function in $V_0^{(d)}$ also belongs to $V_1^{(d)}$, giving rise to a multiresolution analysis.

A generator for $V_0^{(d)}$ can easily be obtained as a suitable translation (necessary to align the breaking points) of

$$\operatorname{rect}^{*(d)}(x) = \begin{cases} \operatorname{rect}(x) = \begin{cases} 1 & \text{if} |x| < 1/2 \\ 0 & \text{otherwise} \end{cases} & \text{if } d = 0 \\ \operatorname{rect}^{*(d-1)} * \operatorname{rect}(x) & \text{if } d > 0 \end{cases}, \tag{59}$$

that is, the rect convolved with itself d times. Function $\operatorname{rect}^{*(d)}$ has compact support but it is not, however, orthogonal to its own translations. It can be orthogonalized with the Fourier trick, but the result has no compact support. For more details about this case, see [3].

3.1 Compactly supported wavelets: Daubechies' wavelets

In the examples above, we found multiresolution analysis whose scaling function had at most two out of the following three desirable characteristics: orthogonality, smoothness, and compact support. Is it possible to find a wavelet that has all these three characteristics? The answer was given by Daubechies. It turns out that imposing all the three characteristics is very demanding and only a small family of wavelets exists.

An easy observation is that if ϕ is orthogonal to its own translations, the coefficients g_i in the two-scale equation can be obtained as

$$g_i = \left\langle \phi, \tau^{i/2} S\phi \right\rangle \tag{60}$$

According to (60) if ϕ has compact support, then g_i has a finite number of coefficients that are different from zero. Since g_i needs to be orthogonal to its even translations, its length (i.e., the number of nonzero coefficients) must be necessarily even [3].

Moreover, if the iteration of operator $\mathcal{O}$ in (39) converges, it is easy to see that g_i has a finite number of coefficients, and then the limit function has compact support. This suggests that it "suffices" to find a finite length sequence g_i that is orthogonal to its own even translation and iterates operator $\mathcal{O}$ to obtain the desired scaling functions. It actually turns out that this can be done, although there are lots of technical details to be taken care of (e.g., about convergence of $\mathcal{O}^k$ and smoothness of the resulting ϕ); see [5] for details.

Every member of the Daubechies family is identified by the length $2N$ of the sequence g_i (remember that the length of g_i is necessarily even). It can be proven that the resulting scaling function has N vanishing moments and its smoothness grows with N; see **Table 1**. See also **Figure 2** that shows the results of the first three iterations of $\mathcal{O}$ (first row), the final scaling function (second row), and the wavelet (third row) of three different Daubechies wavelets.

3.2 Extensions

The construction given above is the original idea of multiresolution analysis. Since the early 1990s, many researchers worked in this field, and many variations and extensions have been introduced. Here we briefly recall those that have more interest in the field of differential equation solutions.

3.2.1 Multiwavelets

Multiwavelets are a generalization of standard multiresolution analysis in the sense that now scaling functions and wavelets are vectors of functions. This means that V_n is not generated by the translations of a single function but from the translations of many functions. Every idea of standard multiresolution analysis can be reformulated without much difficulty in this case, with the most notable difference that the two-scale equation now has vector function and coefficients that are matrices. Multiwavelets can accommodate scaling factors different from two and there is a larger choice for compact support wavelets. See [7] for more details.

N	2	3	4	5	6	7	8	9	10
β_S	1	1.415	1.775	2.096	2.388	2.658	2.914	3.161	3.402
β_H	0.550	0.915	1.275	1.596	1.888	2.158	2.415	2.661	2.902

Table 1.
Hölder β_H and Sobolev β_S regularity exponent of Daubechies' wavelets as function of length 2 N *of* g_i.

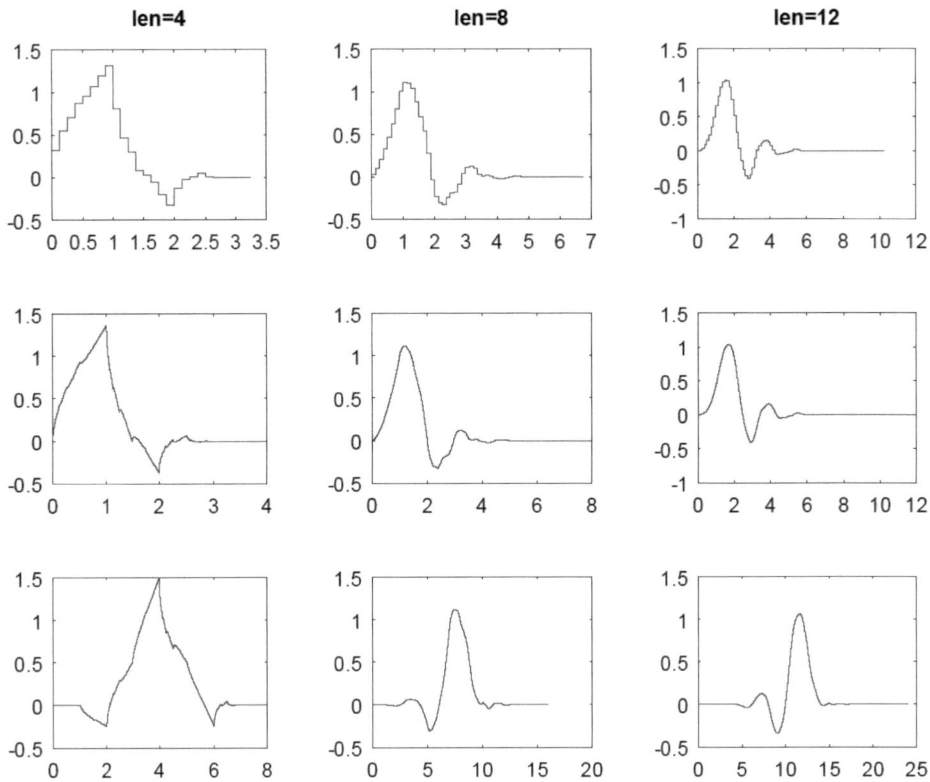

Figure 2.
Daubechies' wavelets. First three iterations of $\mathcal{O}$ (first row), the final scaling function (second row), and the wavelet (third row) of three different Daubechies' wavelets.

3.2.2 Second-generation wavelets

Two-scale Eq. (37) and the resulting filter bank-based procedure work well when the data are sampled on a regular grid and/or the functions of interest are defined on $\mathbb{R}^d$. Since there are many applications that do not satisfy this requirement (e.g., differential equations on general manifolds), the idea of *second-generation wavelet* has been introduced.

The starting point is the so-called lifting form of filter bank (**Figure 1**). It is possible to show that any two-channel filter bank (**Figure 1**) can be implemented as shown in **Figure 3**. In the lifting approach, the input signal is split into odd and even samples by a serial-to-parallel converter. The first branch is filtered, and the result combined with the other branch; the result of this operation is filtered again and combined with the first branch, and this iterated as long as necessary. Filter P is

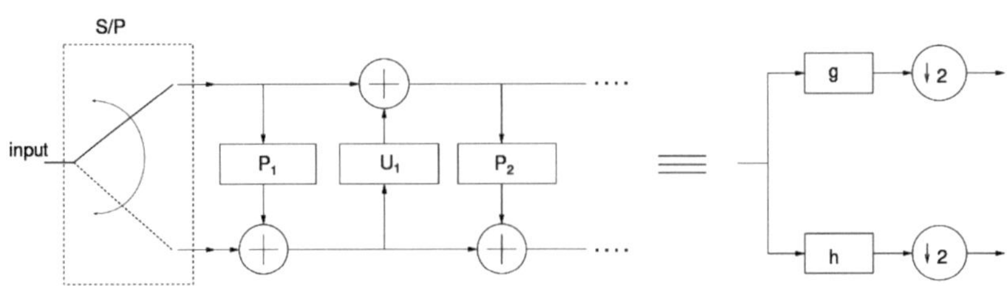

Figure 3.
Lifting implementation of the two-channel filter banks associated with a wavelet analysis.

sometimes called the *prediction* step, and it is interpreted as a filter that predicts the odd samples from the even ones; filter U is sometimes called *update*.

The advantage of this form is that, being similar to the Feistel structure used in cryptography [8], it is exactly invertible even if operations are implemented in fixed-point arithmetic. Actually, the invertibility does not depend on the detail of P_i and U_i; that can be anything, even nonlinear. Another interesting advantage of this predict/update idea is that it does not require a regular domain, allowing to bring the wavelet concept to more general contexts. For example, [9] uses this idea to solve differential equations on the sphere.

4. Some examples of application of wavelets to PDE

By now it should be clear how multiresolution analysis can be applied to differential equation solution: by using scaling functions and/or wavelets as basis functions in approximation (6). All the approaches described in Section 2 can be used with wavelet: collocation, Galerkin PDE method, weighted residual method, meshless methods, etc. Before describing some details of few approaches described in the literature, it is worth to do some general remarks.

What makes wavelet interesting is their multiresolution property and the fact that a wisely chosen wavelet (smooth and/or with many vanishing gradients) has interesting "singularity sensing" properties: in the neighborhood of a singularity (discontinuity, nondifferentiability, etc.), the coefficients decay as a function of scale with a speed that depends on the singularity involved (similar to what Fourier transform does, only on a local level), but away from the singularity, they decade fast [3]. This implies that good approximations can be obtained with few coefficients, using high-resolution decomposition only where it is necessary, reducing the size of the matrices involved in the solution of the PDE. A similar effect can be obtained, for example, in FEM by using a finer mesh around points of large variation. However, using this approach in an adaptive way would require to adjust at running time the mesh, a potentially heavy operation. Wavelets have the potential of employing an adaptive resolution in an easier way. See, for example, [10] for few examples of adaptive techniques employing wavelets.

While orthogonality is considered an important feature in many theoretically wavelet papers, in the context of differential equation solution, it plays a smaller role. The reason is that basis functions enter in the scalar products associated with the various methods via the differential operator $\mathcal{D}$, and it is not guaranteed that $\mathcal{D}$ will preserve orthogonality (that would give rise to many zero entries, that is, sparser matrices).

Actually, orthogonality is preserved if the two basis functions have disjoint support in space (since differential operators do not extend the support) or in frequency (since differential operators are translation-invariant and in frequency they become a product). This suggests that in the context of differential equations, compact support and well-localization in frequency are more important than just orthogonality. In a sense, they represent a "robust" orthogonality condition.

Remark 4.1.

It is true that true compact support in frequency is less common than compact support in space. With the exception of few very special and theoretical cases (e.g., Sinc), the best we can get is a rapid decay in frequency. This means that the scalar product of two basis functions separated in frequency will be maybe very small, but not zero. Nevertheless, even this kind of "almost sparseness" can be exploited.

A general issue with wavelets is that it can be difficult to impose boundary conditions since they have no natural interpolation property that would make boundary condition handling simpler. Another problem that can arise is that many wavelets have no closed-form description, for example, Daubechies wavelets that are described as the result of the iteration of operator $\mathcal{O}$ in (39). This can make their application to PDE more difficult, for example, when computing scalar products involved in weighted residual and other methods.

Finally, another common issue is that most of the known wavelets are defined on a one-dimensional domain, while many physical systems are on a multidimensional domain. The easiest way to create a multidimensional multiresolution analysis by a one-dimensional one is the *separable* (or *tensor product* or *Kronecker product*) approach that create a multidimensional function by the product of several one-dimensional ones, e.g.,

$$\phi_{3D}(x, y, z) = \phi(x)\phi(y)\phi(z) \tag{61}$$

This kind of approach, however, produces "cube-like" wavelets, and their application to FEM schemes based on triangular elements can be difficult.

4.1 Some schemes from the literature

In this section we briefly summarize some interesting wavelet-based schemes that can be found in the literature. As said above, wavelets and scaling functions can be used as basis in the approximation used in collocation, weighted residuals, and other methods.

Wavelets in Galerkin and weighted residual methods bring the advantage of their multiresolution and localization properties while, however, suffering from difficulties in handling complex boundary conditions. Moreover, nonlinear equations can turn out to be difficult to handle. Nevertheless, there have been many successful examples in the application to elliptic, hyperbolic, and parabolic PDE [11–27]. Wavelet-based collocation methods, where wavelet functions are used as shape functions, also registered some success. The advantage of collocation methods is that they are more easily applicable in nonlinear cases [28, 29] and irregular boundary conditions [30]. A collocation method based on second-generation wavelet and lifting is applied to a nonlinear vibration problem in [30, 31].

Much more popular seems to be the application of wavelets to FEM techniques. In this case wavelets or scaling functions are used as shape functions instead of the more traditional polynomials. Daubechies wavelets are particularly popular most probably because of their compact support property. Also of interest is the fact that Daubechies' wavelets can have any number of null moments, making possible the perfect interpolation of polynomials. Some examples of successful application Daubechies wavelets to PDE (mostly mechanical problems) are [32–36]. Of special interest is the proposal of Mitra [37] where wavelet-based FEM is used to transform a wave propagation problem into ordinary differential equations that are successively solved.

Another popular solution for wavelet-based FEM is the wavelets based on spline spaces. Although spline bases cannot have both compact support and orthogonality, in differential equations, as explained above, we gladly give up on orthogonality if we can get compact support and smoothness. Another important advantage of splines is that a simple closed-form expression is known. Examples of spline applications can be found in [38–40]. Of special interest is the application of Hermite cubic splines (HCS), a kind of multiwavelet [41] that shows promise in handling in a numerically robust way boundary conditions. The HCS is a multiwavelet with four smooth (twice differentiable) components defined on interval $[0, 1]$. Some examples of application can be found in [42–44]. A problem with the application of wavelets to FEM techniques is the difficulty to adapt the wavelet construction to complex meshes. In this case the use of second-generation wavelet based on an extension of the lifting idea has attracted some attention [45–47].

Wavelets have attracted some interest also in the context of meshless methods [9, 48, 49]. Of some interest for the special problem is [9] that uses a wavelet approach to implement a meshless solver for differential equations defined on the sphere. A problem with applying wavelets to generic manifolds like a sphere is that it is not clear what a "rescaling by 2" should mean for a manifold that is not a Euclidean space. The idea used in [9] is to use a so-called *diffusion wavelet* where the dilation is replaced by a *diffusion operator* that looks like a kind of "low-pass filtering" that smear out the details; see [9] for the precise definition.

5. Conclusions

This chapter introduced the reader to the field of applying wavelets to the numerical solution of differential equations. Both wavelets and differential equations are research fields with many applications, contributions, and results. Their combination gives rise to wide varieties of methods, each one suited for specific applications. By looking at the literature, we can see that wavelets can be a very powerful tool for solving PDE especially because of their multiresolution nature that allows to optimize the level of detail where it is needed. Wavelets, however, are not a silver bullet for all problems either, since they can have some characteristics (multidimensional construction via tensor product, nonexistence of a closed-form expression, difficulty in handling some boundary conditions, etc.) that can make their application not trivial in some cases. We can say that this is a field where, more than ever, no *single solution fits all* and that every practitioner needs to find the solution specific for the problem at hand using knowledge in both fields and some ingenuity.

Author details

Riccardo Bernardini
University of Udine, Udine, Italy

*Address all correspondence to: riccardo.bernardini@uniud.it

References

[1] Mallat S. Multiresolution approximations and wavelet orthonormal bases of $L^2(R)$. Transactions of the American Mathematical Society. 1989;**315**:69-87

[2] Meyer Y. Ondelettes, Vol. 1 of Ondelettes et Opérateurs. Paris: Hermann; 1990

[3] Vetterli M, Kovačević J. Wavelets and Subband Coding. Signal Processing. Englewood Cliffs, NJ: Prentice-Hall; 1995

[4] Huerta A, Belytschko T, Fernández-Méndez S, Rabczuk T, Zhuang X, Arroyo M. Meshfree Methods. In: Stein E, Borst R, Hughes TJ, editors. Encyclopedia of Computational Mechanics. 2nd ed. 2017. DOI: 10.1002/9781119176817.ecm2005

[5] Daubechies I. Ten Lectures on Wavelets. Philadelphia: SIAM; 1992

[6] Mallat S, Hwang WL. Singularity detection and processing with wavelets. IEEE Transactions on Information Theory. 1992;**38**:617-643

[7] Keinert F. Studies in advanced mathematics. In: Wavelets and Multiwavelets. Chapman and Hall/CRC; 2003

[8] Schneier B. Applied Cryptography: Protocols, Algorithms, and Source Code in C. Hoboken, New Jersey: John Wiley and Sons; 1994

[9] Goyal K, Mehra M. An adaptive meshfree diffusion wavelet method for partial differential equations on the sphere. Journal of Computational Physics. 2014;**272**:747-771

[10] Li B, Chen X. Wavelet-based numerical analysis: A review and classification. Finite Elements in Analysis and Design. 2014;**81**:14-31

[11] Hussain NEA. 2D wavelets Galerkin method for the computation of EM field on seafloor excited by a point source. International Journal of Applied Electromagnetics and Mechanics. 2017; **53**(4):631-644

[12] Qian S, Weiss J. Wavelets and the numerical solution of partial differential equations. Journal of Computational Physics. 1993:155-175

[13] Amaratunga K, Williams JR. Wavelet based green's function approach to 2d pdes. Engineering Computations. 1993;**10**(4):349-367

[14] Ho S, Yang S. Wavelet-galerkin method for solving parabolic equations in finite domains. Finite Elements in Analysis and Design. 2001;**37**(12): 1023-1037

[15] AL-Qassab M, Nair S. Wavelet-galerkin method for free vibrations of elastic cable. Journal of Engineering Mechanics. 2003;**129**(3):350-357

[16] Monasse P, Perrier V. Orthonormal wavelet bases adapted for partial differential equations with boundary conditions. SIAM Journal on Mathematical Analysis. 1998;**29**(4): 1040-1065

[17] Nastos CV, Theodosiou TC, Rekatsinas CS, Saravanos DA. A 2d daubechies finite wavelet domain method for transient wave response analysis in shear deformable laminated composite plates. Computational Mechanics. 2018;**62**:1187-1198

[18] Dahmen W, Kunoth A, Urban K. A wavelet galerkin method for the stokes equations. Computing. 1996;**56**:259-301

[19] Yang S, Ni G, Ho SL, Machado JM, Rahman MA, Wong HC. Wavelet-galerkin method for computations of electromagnetic fields-computation of connection coefficients. IEEE Transactions on Magnetics. 2000;**36**: 644-648

[20] Yongping S, Pu L, Rufu H. A wavelet-galerkin method for the simulation of torsion mems devices under the effect of squeeze film damping. In: 2011 International Conference on Electric Information and Control Engineering; 2011. pp. 5671-5674

[21] Yang S, Ni G, Cardoso JR, Ho SL, Machado JM. A combined wavelet-element free galerkin method for numerical calculations of electromagnetic fields. IEEE Transactions on Magnetics. 2003;**39**: 1413-1416

[22] Liu Y, Liu Y, Ding K. A new coupling technique for the combination of wavelet-galerkin method with finite element method in solids and structures. International Journal for Numerical Methods in Engineering. **112**(10): 1295-1322

[23] Venini P, Morana P. An adaptive wavelet-galerkin method for an elastic-plastic-damage constitutive model: 1d problem. Computer Methods in Applied Mechanics and Engineering. 2001; **190**(42):5619-5638

[24] Sannomaru S, Tanaka S, ichiro Yoshida K, Bui TQ, Okazawa S, Hagihara S. Treatment of dirichlet-type boundary conditions in the spline-based wavelet galerkin method employing multiple point constraints. Applied Mathematical Modelling. 2017;**43**: 592-610

[25] Yang Z, Liao S. On the generalized wavelet-galerkin method. Journal of Computational and Applied Mathematics. 2018;**331**:178-195

[26] Bu H, Wang D, Zhou P, Zhu H. An improved wavelet–galerkin method for dynamic response reconstruction and parameter identification of shear-type frames. Journal of Sound and Vibration. 2018;**419**:140-157

[27] Choudhury A, Deka R. Wavelet–galerkin solutions of one dimensional elliptic problems. Applied Mathematical Modelling. 2010;**34**(7):1939-1951

[28] Zhang T, Tian Y-C, Tadé MO. Wavelet-based collocation method for stiff systems in process engineering. Journal of Mathematical Chemistry. 2008;**44**:501-513

[29] Lepik Ü. Haar wavelet method for solving stiff differential equations. In: Mathematical Modelling and Analysis. Vilnius, Lithuania: Vilnius Gediminas Technical University; 2009;**14**:467-481

[30] Vasilyev OV, Bowman C. Second-generation wavelet collocation method for the solution of partial differential equations. Journal of Computational Physics. 2000;**165**(2):660-693

[31] Vasilyev OV, Kevlahan NK-R. An adaptive multilevel wavelet collocation method for elliptic problems. Journal of Computational Physics. 2005;**206**(2): 412-431

[32] Adigun BJ, Buchan AG, Adam A, Dargaville S, Goffin MA, Pain CC. A haar wavelet method for angularly discretising the boltzmann transport equation. Progress in Nuclear Energy. 2018;**108**:295-309

[33] Patton RD, Marks PC. One-dimensional finite elements based on the daubechies family of wavelets. AIAA Journal. 1996;**34**:1696-1698

[34] Ma J, Xue J, Yang S, He Z. A study of the construction and application of a daubechies wavelet-based beam element. Finite Elements in Analysis and Design. 2003;**39**(10):965-975

[35] Li B, Hongrui C, He Z. The construction of one-dimensional Daubechies wavelet-based finite elements for structural response analysis. Journal of Vibroengineering. 2011;**13**:729-738

[36] He W-Y, Wang Y, Zhu S. Adaptive reconstruction of a dynamic force using multiscale wavelet shape functions. Shock and Vibration. 2018;**2018**:8213105

[37] Mitra M, Gopalakrishnan S. Wave propagation analysis in anisotropic plate using wavelet spectral element approach. Journal of Applied Mechanics. 2008;**75**(1):014504

[38] Han J-G, Ren W-X, Huang Y. A spline wavelet finite-element method in structural mechanics. International Journal for Numerical Methods in Engineering. **66**(1):166-190

[39] Chen WH, Wu CW. A spline wavelets element method for frame structures vibration. Computational Mechanics. 1995;**16**:11-21

[40] Pian THH, Chen D-P. Alternative ways for formulation of hybrid stress elements. International Journal for Numerical Methods in Engineering. 1982;**18**(11):1679-1684

[41] Jia R-Q, Liu S-T. Wavelet bases of hermite cubic splines on the interval. Advances in Computational Mathematics. 2006;**25**:23-39

[42] Xiang J-w, Chen X-f, Li X-k. Numerical solution of poisson equation with wavelet bases of hermite cubic splines on the interval. Applied Mathematics and Mechanics. 2009;**30**:1325

[43] Chen X, Xiang J. Solving diffusion equation using wavelet method. Applied Mathematics and Computation. 2011; **217**(13):6426-6432

[44] Xiang J, Wang Y, Jiang Z, Long J, Ma G. Numerical simulation of plane crack using hermite cubic spline wavelet. CMES: Computer Modeling in Engineering and Sciences. 2012;**88**:1-16

[45] He Y, Chen X, Xiang J, He Z. Adaptive multiresolution finite element method based on second generation wavelets. Finite Elements in Analysis and Design. 2007;**43**(6):566-579

[46] He Y, Chen X, Xiang J, He Z. Multiresolution analysis for finite element method using interpolating wavelet and lifting scheme. Communications in Numerical Methods in Engineering;**24**(11):1045-1066

[47] Sudarshan R, D'Heedene S, Amaratunga K. A multiresolution finite element method using second generation hermite multiwavelets. In: Bathe K, editor. Computational Fluid and Solid Mechanics 2003. Oxford: Elsevier Science Ltd; 2003. pp. 2135-2140

[48] Liu Y, Liu Y, Cen Z. Daubechies wavelet meshless method for 2-d elastic problems. Tsinghua Science and Technology. 2008;**13**:605-608

[49] Yousefi MR, Jafari R, Moghaddam HA. A combined wavelet based mesh free method for solving the forward problem in electrical impedance tomography. In: 2012 IEEE International Symposium on Medical Measurements and Applications Proceedings; 2012. pp. 1-4

Section 2

Real World Applications

Chapter 4

DWT-Based Data Hiding Technique for Videos Ownership Protection

Farhan Al-Enizi and Awad Al-Asmari

Abstract

This chapter proposes a wavelet data hiding scheme for video authentication and ownership protection. A watermark in the shape of a logo image will be hidden. In this research, a discrete wavelet transform (DWT) process is implemented using orthonormal filter banks, where the Y components of the YUV color space of the video frames are decomposed using DWT, and a watermark is inserted in one or more of the resulting sub-bands in a way that is fully controlled by the owner. Then, the watermarked video is reconstructed. The filters used for the DWT decompositions are randomly generated to increase the security of the algorithm. An enhanced detection technique is developed to increase the reliability of the system. The overall robustness of this scheme is measured when common attacks are applied to the test videos. Moreover, the proposed algorithm is used with the high-efficiency video coding (HEVC) technique to examine the whole performance. Furthermore, a selective denoising filter is built to eliminate the effect of the noise. The simulation results show that the proposed algorithm achieves well under both the visual and the metric tests. Moreover, it performed well against intentional and unintentional attacks. The average normalized correlation achieved is 97%, while the mean peak signal-to-noise ratio (PSNR) is 45 dB.

Keywords: watermarking, filter banks, data hiding, pyramid transform, video coding, video attacks

1. Introduction

The digital age brought new technologies and services to people, industry, and governments. The digitization process covered all types of information being exchanged especially images and videos. The exchange of these forms of multimedia became faster and easier with the new communication network capabilities. On the other hand, that made it easier to steal or use these forms of multimedia illegally. Those concerns opened new horizons in the field of multimedia security especially data hiding [1]. Data hiding has many types; a useful classification is to divide data hiding into watermarking and steganography. New techniques used in video processing brought new challenges and difficulties to the data hiding methods. Moreover, the new compression techniques especially the high-efficiency video coding (HEVC) or H.265 are added to these challenges and difficulties to have

 IntechOpen

reliable, secure, and robust data hiding methods [2, 3]. Various watermarking schemes that use different techniques have been proposed over the years [4–9]. To be effective, a watermark must be imperceptible within its host, extracted with ease by the owner, and robust in the face of both intentional and unintentional distortions [7, 10, 11]. In specific, discrete wavelet transform (DWT) has wide applications in the different areas of image and video processes such as compression, noise reduction, and watermarking [12]; this is attributed to its characteristics in space-frequency localization, multi-resolution representation, and superior human visual system (HVS) modeling [5]. The robustness is a very important aspect in data hiding or watermarking. To achieve the highest levels of robustness, new methods and techniques should be introduced and optimized at both the sender and receiver sides. Furthermore, the detection process should be enhanced to meet these requirements.

In this research, a video watermarking process that depends on the discrete wavelet decompositions will be developed. Moreover, the detection process will be enhanced through statistical derivations. The security will be maintained through the adoption of random filter banks, the study of the motion and motionless scenes in the video frames, and the spread spectrum generation of the watermarks. The overall technique has to meet the requirements of visual quality, security, robustness, and computational complexity.

2. Proposed watermarking technique

In this section, we introduce our digital video watermarking technique for the purpose of authentication and ownership protection. The proposed technique is aimed at achieving reasonable degrees of robustness, visual quality, and security. The embedding technique involves two stages: first, a decomposition process and then a hiding process. The watermark can be any binary sequence; normally a binary image of a specific size is used. The encoded videos can be in any color space; in our case, YUV space is used. It is possible to perform the hiding process in any of the three components: Y, U, and V. In this work, the luminance Y frames are used as host images for the data hiding process; that is, the hiding of the watermark will be performed in one or more of the sub-bands that result from the discrete wavelet analysis process. Choosing the wavelet filters is an important aspect in the efficiency of the reconstruction process; special types of filters are the randomly generated orthonormal filter banks [13]. These filter banks can be generated randomly depending on the generating polynomials; hence, by generating random numbers for the polynomial coefficients, it is possible to build multiple filter banks that are used for the different stages of our decomposition processes. The orthonormal analysis and synthesis filters can be generated in different ways; for our technique, having large side-lobes is preferred. This enables us to hide more energy in the medium frequencies of the image; in doing so, we construct a more robust way that can counteract the effects of different image processes, which take place intentionally or unintentionally over the course of the handling process. Each filter bank that is generated is used for one level of the DWT analysis and synthesis processes. Moreover, the number of the levels and the structure that is followed during the analysis process are controlled by the owner. It is well known to the image processing community that the medium-frequency bands are preferred for hiding. This will avoid hiding in the lower-frequency bands where most of the energy is concentrated and the higher-frequency bands where the possibility of losing the data is high due to compression processes. Furthermore, the possibility of using more than one sub-band rather than a single sub-band is there; this method is useful in having a robust method against the nonlinear collusion attack.

There are many scenarios that can be followed for the embedding process; one of them is to embed the data which is our binary watermark using a generated pseudorandom sequence [14]. This method depends mainly on doing the watermarking process by converting the original binary watermark image Q to some sort of a binary sequence S of a specific length M; in this case, the data pixels are given the value +1, whereas the background pixels are given the value −1. Furthermore, a pseudorandom sequence P of the same length M as our watermark sequence is generated using a secret key; likewise, this sequence is represented by values that are either +1 or −1. The DWT coefficients of the decomposed sub-bands that will be used for the hiding process are represented as a matrix Q_1 of the same size as our watermark. Moreover, it can be written as a vector T of length M. The binary watermark is hidden into this vector T, and that will result in a new vector that is called T' according to the rule that is shown in this equation:

$$t_i' = t_i + \alpha * p_i * s_i, for\ i = 1, 2 \dots M \tag{1}$$

where α is a numerical factor which represents a weighting constant that determines the strength of the processed watermark. This number is chosen in such a way to offer a trade-off between the required robustness and the acceptable visual quality. Moreover, choosing this weighting factor should take into consideration many elements in image processing techniques such as the compression standard that is used and its intensity, the smooth features or the textures that are there in the image, and the algorithm that is followed when doing the detection process. Furthermore, how much energy content is there in the wavelet sub-bands must be considered at the hiding stage. One way to get the numerical magnitude factor is to have a comparison process between the energy of the original coefficients of the host DWT sub-band Q_1 and energy content of the original watermark image Q elements according to this empirical formula:

$$\alpha = 2 * \sqrt{\frac{E(Q_1)}{E(Q)}} \tag{2}$$

where $E(Q_1)$ represents the energy content of the original wavelet coefficients, while $E(Q)$ represents the energy content of the watermark image Q; the energy was computed by taking the sum of the squared elements. The manipulated wavelet coefficients according to our hiding process are used then depending on their respective locations to reconstruct and build the watermarked image frame. The overall hiding process of a binary watermark for a Y frame is shown in **Figure 1**. It is clear from this figure, and this, in fact, depends on the decomposition structure that is followed that the low-low (LL) frequency area of the decomposed image is not used for our embedding process. This area or band is called the decimated image normally, and it results in both the pyramidal and DWT decompositions. It is clear that this band or image has most of the information or energy of the original image frame; the other images in other bands are normally called the error images, and they have lower energy content. In fact, they represent other bands depending on the analysis filters which are the low-high (LH), high-low (HL), and high-high (HH) bands. These bands offer better places for the hiding process.

The watermark, which is primarily a binary image, can be embedded in any of the frames of the host video; moreover, the frames can be chosen in a fully controlled selective way. The degree of randomness that is achieved is up to the user

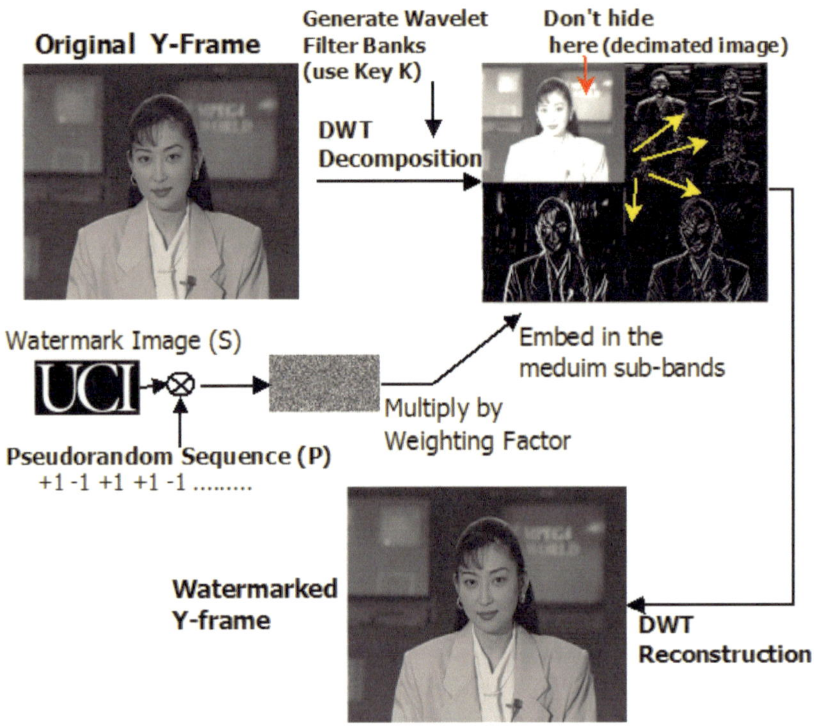

Figure 1.
The block diagram of the proposed watermarking method.

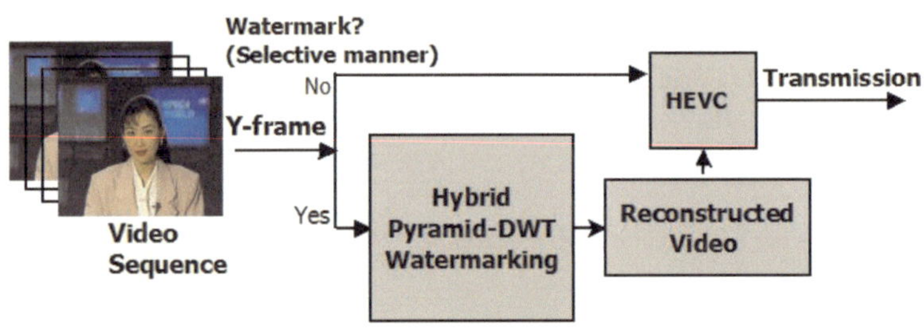

Figure 2.
The block diagram of the watermarking process with the application of HEVC process.

who is the sole owner. Furthermore, the security of the system depends partially as well on the degree of the randomness of the pseudorandom sequence that is used in the encoding process. On the other hand, the Y components of the color space were chosen intentionally because they have higher resolution and therefore higher hiding capacity, but we have to keep in mind the fact that the U and V components likewise can be used. As we mentioned in the introduction, our techniques will be used when the HEVC process is applied; **Figure 2** shows the proposed hiding process when the HEVC or H.265 process is applied to the video that is watermarked.

2.1 1D discrete Fourier transform

When designing a robust and dependable embedding system, security concerns always come to the forefront. Hiding the same watermark in a repetitive manner to each and every frame of the host video may cause a problem of maintaining the statistical invisibility, which is an important condition of every security system [15]. Moreover, applying independent watermarks to each and every of these frames also causes a security problem if these frames have few or no motion areas inside them; these motionless regions in successive video frames may be statistically compared or averaged to remove independent watermarks. Attacks of such kinds are normally called collusion attacks. The inter-frame collusion attacks, for instance, exploit the repetition in the video frames and their scenes or in the watermarks themselves to produce a false copy of the video that does not have any watermarks; these attacks can be divided into watermark estimation remodulation (WER) attack and frame temporal filtering (FTF) attack [16]. Classifying the video frames according to the amount of motion in them is useful in this regard. The motion in videos is a relative one, since most of the videos have motion, but what interest us here are the amount of this motion, how fast this motion is, the relative motion with respect to the surroundings, and the distribution of this motion across the frames. Most of the video compression techniques use inter-frame motion estimations to encode the frames; however, there are other methods that can be used to detect static and dynamic scenes in videos. One method can be built depending on the *1D discrete Fourier transform* (DFT). The 1D DFT in temporal direction performs a transformation process of a group of pictures (GOPs) into a temporal frequency domain; in the resulting domain, both the video frames spatial and temporal frequency information exist in the same resulting frame. Higher frequencies are a reflection of the fast motion from one frame to other frames [17]. The 1D DFT of a video $f(x,y,t)$ that has a specific size of MxNxT, in which MxN is the size of each of the video frames and T is the number of the video frames that are grouped in one GOP, is given by

$$F(u,v,\tau) = \sum_{t=0}^{T-1} f(x,y,t) e^{-j2\Pi(t\tau/T)} \tag{3}$$

where u and v represent the spatial domain of the video frames, while τ represents the temporal domain of these frames. Normally the GOPs are taken as five frames or a close number. Depending on that, a group of the so-called spatiotemporal frames can be constructed for the Foreman video. Twenty-five frames of the Foreman video were transformed using this method of the 1D DFT, and since the DFT is a symmetric process in one GOP, so it is logical to show only the first spatiotemporal frame of each of those groups of pictures. **Figure 3** shows the first frame of the Foreman video, while **Figure 4** shows the 5 temporal frames of this

Figure 3.
The first frame of the Foreman video.

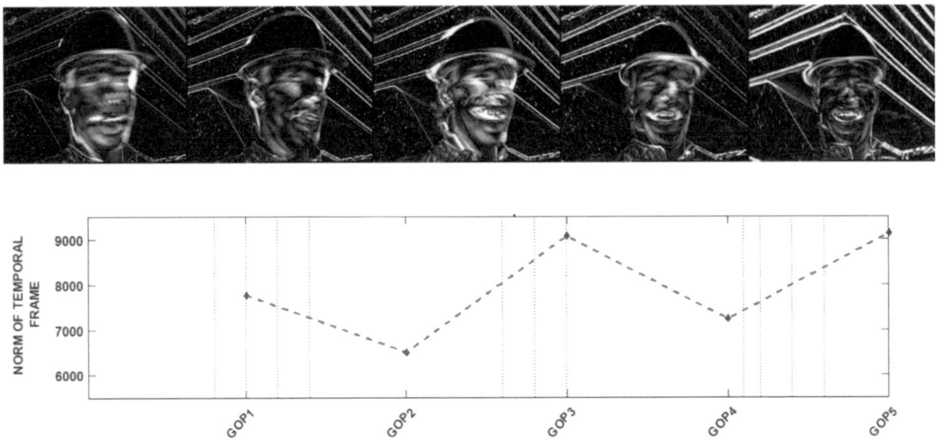

Figure 4.
The 1D DFT of 5 GOPs of 25 frames of Foreman video and their corresponding norms.

video that were evaluated from the original 25 frames; their norms are shown as well. The edges that are seen in these frames correspond to high frequencies which reflect the motion in temporal domain and how this motion in each frame is distributed; furthermore, the values of the evaluated norms reflect how much and how fast the motion in each GOP is. For instance, the intensity of the edges shows the movement of the head and the relative motion with respect to the building; it can be seen that the background has some motion that corresponds to a moving camera which is exactly the case here.

Depending on the previous analysis of the videos using the 1D DFT and the classification of the video frames into dynamic and static frames, a significant enhancement can be added to the hiding process in terms of both security and reliability. Using this analysis, different binary watermarks will be embedded in motion frames, and the same binary watermark will be embedded in motionless ones. In fact, since we need to have some repetition of the watermark to enhance the detection process, this method helps us without weakening our algorithm due to statistical estimation methods that are used in steganalysis, for instance; moreover repeating the watermark in motionless frames increases the cohesion of the watermarked video sequence. Furthermore, the bands being used are not confined to the high-frequency ones; the effect of averaging and collusion attacks is reduced as well. Using 1D DFT to establish motion information is not the only way that can be used; 3D DWT, for instance, can be used to construct spatiotemporal components of videos frames. Choosing the proper method to determine motion in frames depends in the first place on the application and other elements such as computational complexity. Since we are only looking for a method to estimate motion but not in a strict and precise way, using 1D DFT meets our needs at this stage.

3. Watermark detection process

The extraction process depends mainly on the hiding process, and so we are performing a reverse process. This is a blind watermarking method; hence, knowing the original watermark image is not a requirement, but, still, knowing the reconstruction synthesis filter banks and the generated pseudorandom sequence is required to extract our hidden watermark. To get the hidden watermark, a prediction and estimation process of the original values of the pixels is required [14]. This process should also take into account that different types of processing will take place such as the lossy compression, the additive noise, and the geometrical operations; this, in turn, renders the detection process a challenging one. Furthermore, security concerns arise as a critical point; this is reflected in the attempts of attackers to know or destroy the hidden watermark. To cope with these difficulties, an enhanced detection and estimation process is developed.

A noise elimination method can be developed to estimate the original pixels; to do that, the extracted coefficients can be smoothened using a spatial convolution mask of size 5x5. In fact the 5x5 mask gave higher performance than the 3x3 mask when our videos were subjected to noise and compression. Moreover, the selective denoising filter which is presented in Section 4 gave good results in removing the noise and smoothening the extracted image. Using a subtraction operation as opposed to the additional one in Section 2 and setting the positive and negative values to +1 and −1, respectively, a coarse version of the watermark can be extracted. The enhanced detection process is then set to use multiple extracted watermarks, which were embedded randomly in different video frames in the first place, for our final estimation process. It was shown in the previous section that either the same watermark or multiple watermarks can be used depending on the changes in the scenes; moreover it was shown that the 1D DFT is helpful in determining these changes. This means that we are not sacrificing the security when using the same watermark in a random manner; on the contrary we are increasing robustness against detection or manipulating attempts. Let us assume that the extracted watermarks are grouped in a set $W = w_1, w_2, \ldots w_n$. To choose the set of watermarks that can be used in the final estimation process, cross-correlation test can be performed between every two extracted watermarks w_i and w_j. The normalized cross-correlation coefficient between two matrices A and B is given according to the following equation:

$$R = \frac{\sum_m \sum_n \left(A_{mn} - \overline{A}\right) \ \left(B_{mn} - \overline{B}\right)}{\sqrt{\left(\sum_m \sum_n \left(A_{mn} - \overline{A}\right)^2 \ \right)\left(\sum_m \sum_n \left(B_{mn} - \overline{B}\right)^2\right)}} \tag{4}$$

where $\overline{A}$ and $\overline{B}$ are the means of A and B, respectively. The attacks that the videos are subjected to are of different natures and scopes; they can be divided into geometrical, statistical, additive noise, etc. Hence, the watermarks that are extracted can be talked about as noisy versions of the originally hidden ones or, equivalently, noisy signals. The cross-correlation test gives good indication of the similarity between two signals, and this can be applied to our extracted watermarks which are expected to have some sort of similarity. Depending on this statistical analysis, it is possible to establish a new set of extracted watermarks W_1.

Depending on the resulting cross-correlation value, it is possible to get the set of coefficients that can be used in our final decision-making process. Hence, the final watermark set W_1 can be established. On the other side, if the correlation value was low, that means that the coefficients are so corrupted, and therefore they will be excluded from our final set. This cross-correlation process can be seen in **Figure 5** where **Figure 5(a)** shows a plot of the cross-correlation matrix between two sets of coefficients that are highly correlated, and that means they can be included in our final set, while **Figure 5(b)** shows the opposite of that, where these coefficients are corrupted. To establish a good estimation process, a threshold value should be defined for the cross-correlation value, and the decision can be done accordingly.

Since the cross-correlation between binary images is a measure of similarity between these images, this tells us that flipping the value of any pixel will reduce this similarity. If $w_i \in W$, then a cross-correlation process is performed between w_i

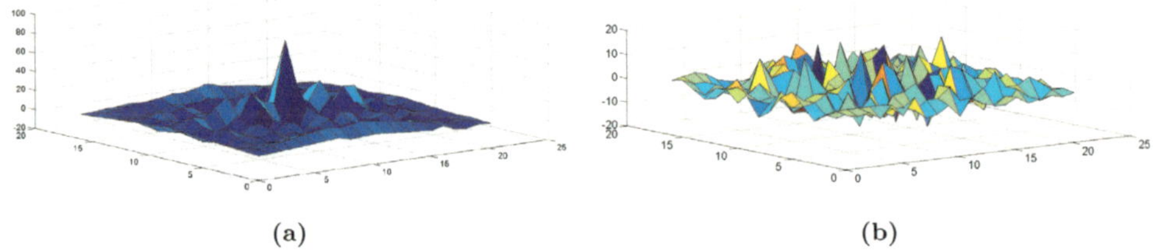

Figure 5.
3D plots of the cross-correlation matrices of two extracted watermarks.

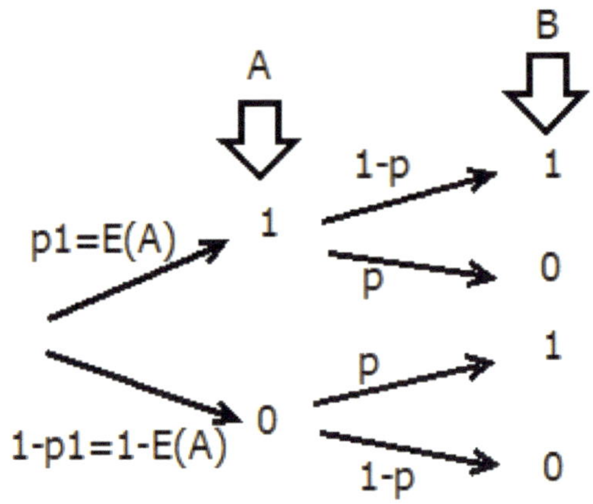

Figure 6.
Expected values of the input and output binary images.

and all the other extracted watermarks in the set; then the average cross-correlation parameter is evaluated. The same process is done for all the watermarks in the set. A set that includes each extracted watermark and its corresponding average correlation value is established. Then, by establishing a threshold value h for the average cross-correlations, the extracted watermarks that do not achieve the threshold test are excluded from the new set W_1. The final extracted watermark w_e can be evaluated by performing an averaging process on the watermarks in the set W_1, where

$$w_e = Ave\{W_1\} \tag{5}$$

Doing an averaging process is attributed to the fact that binary sets follow specific statistical pattern. The correlation coefficient R between any two arbitrary matrices A and B is given in Equation 4; the mean value of a binary image A is at the same time the expected value of A or E(A). Assuming that at the input, the probability of 1 is p_1 and that the probability of flipping of the value is p as shown in **Figure 6**, then

$$\overline{A} = E(A) = p_1 \tag{6}$$

Moreover, the probability of having 1 at the output $\overline{B} = p_1 * (1-p) + \left(1-p_1\right) * p$ and by taking Equation 6 into consideration, this equation can be rewritten as

$$\overline{B} = E(B) = E(A) + (1 - 2 * E(A)) * p \tag{7}$$

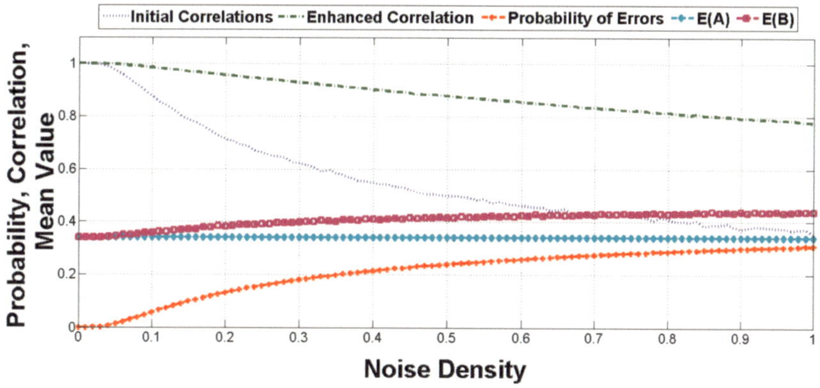

Figure 7.
The enhanced correlations vs. noise density.

Assuming that we are using a specific binary watermark, then the input matrix A is constant during our watermarking process. This means that in the above equation, the flipping probability of the pixels p is the sole variable. Moreover, by having a comparison between Eqs. (4) and (7), it can be seen that the correlation between the two matrices A and B is dependent on the flipping probability of the pixels; hence, the flipping effect is reduced somehow by doing the averaging process of the extracted watermarks. An enhanced version can be built as far as p is not equal to the value 0.5 which corresponds to a unity entropy value. To demonstrate this analogy, **Figure 7** shows the changes in these parameters when a random binary watermark is subjected to Gaussian noise with zero mean and different variances; in this figure, the variance of noise is represented by the term density for the illustration and clarification purposes.

4. Noise removal selective filter

One of the challenging aspects in video encoding and watermarking is the additive noise that results in distorted video streams. The nature of the additive noise depends primarily on the source of this noise. Not only the additive noise tends to distort the visual quality of the video in question, but it also has its noticeable impacts on the watermarking process. One type of noises that is common in video processing techniques is the salt-and-pepper (S&P) noise. This type of noise could be added to the video frames during the transmission process when the communication channels, in a sense, are noisy, or it could be a result of the hardware-generated errors during the encoding and decoding processes. Removing the noise without disturbing the watermarking process on the one hand and preserving the visual qualities on the other hand is a challenging process. As far as the watermarking process is concerned, it is useful to check the effects of both the additive noise and the removal process on our data hiding process. Many methods were proposed to eliminate the noise or enhance the visual appearance of the images [18, 19]; these methods depend mainly on the idea of median filters. The normal median filters, for example, which are used to eliminate the salt-and-pepper noise in images, do in fact filter the whole image regardless of the presence or absence of the noise in a certain area. This process reduces the original resolution of the image to a great extent in such a way that the qualities of high-definition (HD) videos are lost. This means that our watermarking process would not achieve the visual quality condition. In this research, a noise detection process that depends on the absolute differences between a pixel aij and its surrounding pixels is proposed. In order to enhance the detection process, the variance of the pixels in the surrounding window is calculated. This step is important because of false detections, especially at edged and textured details of the image where the absolute difference value could be high, while the region is noise-free. This method takes into account the fact that such variances are dramatically high at these locations. However, this is not the case around noisy pixels in general where some sort of consistency is there. The proposed method for noise detection and elimination process involves the following steps:

1. For each pixel a_{ij}, a sub-window of size 3x3 around this pixel is taken.

2. The absolute differences between the pixel a_{ij} and the surrounding pixels are calculated.

3. The arithmetic mean (AM) of the calculated differences for a given pixel a_{ij} is computed. The AM is then compared with a threshold value t to detect whether the pixel a_{ij} is informative or corruptive.

4. The 3x3 pixel window is converted to an array, and then it will be arranged in an ascending order. The largest and the smallest values will be eliminated. This will help in removing other noisy pixels in the surrounding window. The resulting array will be denoted L. The variance of the pixels in the array L is computed and denoted as V.

5. A comparison will be performed between AM and V on one side and their respective thresholds on the other side:

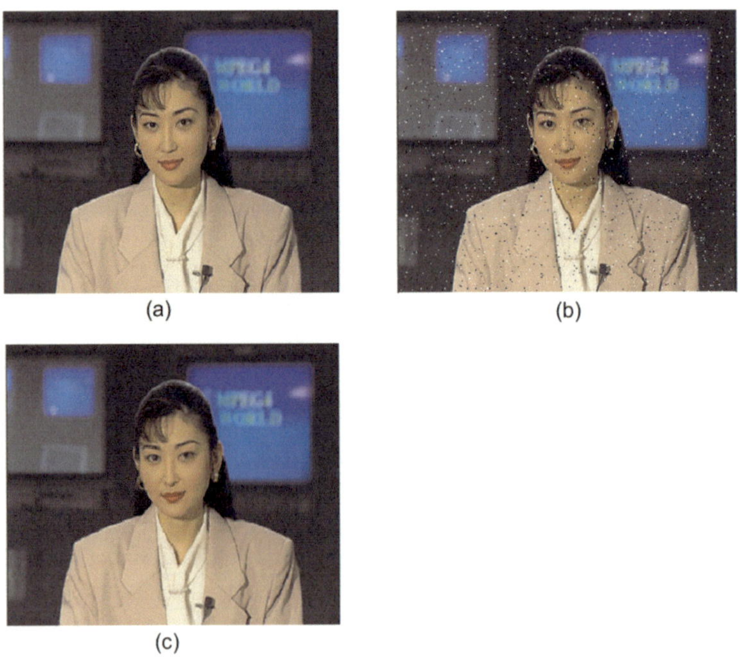

Figure 8.
(a) Original Akiyo frame; (b) 2% S&P noisy Akiyo frame; and (c) the denoised frame.

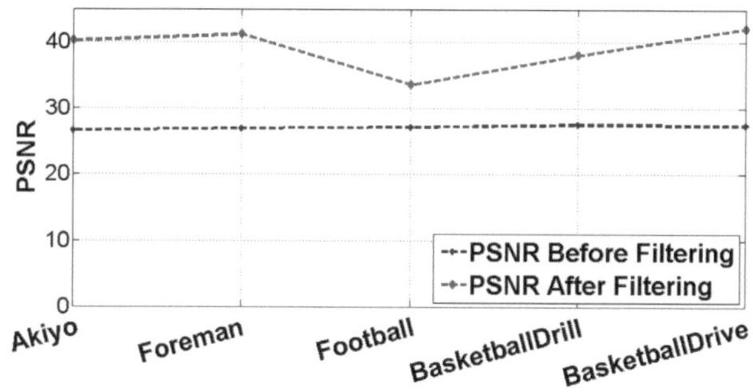

Figure 9.
PSNRs of standard videos before and after denoising process.

- If *AM* is greater than *t* and *V* is less than the variance threshold, then do the elimination process by replacing the noisy pixel by the median of the surrounding pixels in the window.

- Otherwise, do nothing. In this case, either there is no noise, or the pixel in question is on one of the edges of the image, and nothing should be done accordingly. Smooth and textured images perform differently with respect to noise; it is easier to remove noise, specifically salt-and-pepper noise from smooth images.

The arithmetic mean (AM) threshold is a user-defined value between the minimum and maximum pixel values (0.255) which are used to distinguish an informative pixel from a noisy one. On the other hand, the variance *V* can take larger values, and its threshold value can be determined accordingly. In fact, its value depends on the images themselves whether they were textured or smooth ones. The original Akiyo frame, a noisy version of this frame with salt-and-pepper noise of 2% density, and the same frame after the denoising process are shown in **Figure 8**. **Figure 9** shows the peak signal-to-noise ratio (PSNR) values of the noisy and denoised versions of the standard videos: Foreman, Akiyo, Football, BasketballDrill, and BasketballDrive.

5. Experimental results

In this section we demonstrate the performance of our algorithm using our proposed method on different standard videos with and without HEVC process, under different attacks. Furthermore, it will be compared with the method in [20]. Watermarked and unwatermarked versions of a frame of BasketballDrill video (832×480 pixels) are shown in **Figure 10**. The embedded and extracted watermarks of size 15×26 are shown in **Figure 11**; in fact, they are enlarged for illustration purposes.

Our algorithm performance will be evaluated in terms of PSNR between the original and the watermarked videos and the normalized correlation (NC) between the original and the extracted watermarks for the standard videos: Foreman, Akiyo, Football, BasketballDrill, and BasketballDrive. For the CIF (352×288) videos, a 9×11 watermark was used, while for the other two videos, the watermark in **Figure 11(a)** was used. In these tests, 100 frames were watermarked. **Figure 12** shows the NC of the extraction process; moreover, **Figure 13** shows the enhanced

Figure 10.
The first frame of (a) original BasketballDrill frame and (b) watermarked BasketballDrill frame.

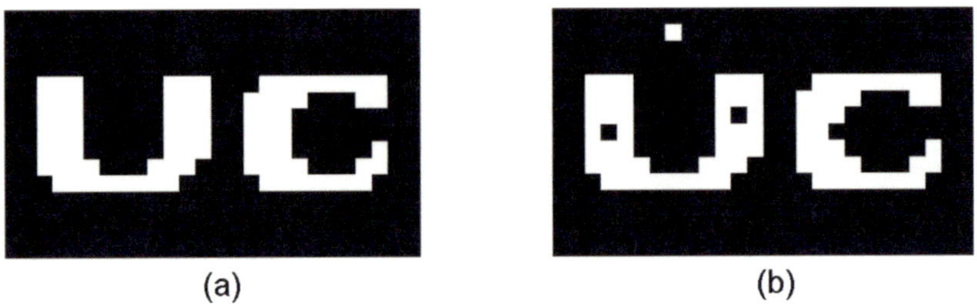

Figure 11.
(a) Original watermark and (b) recovered watermark.

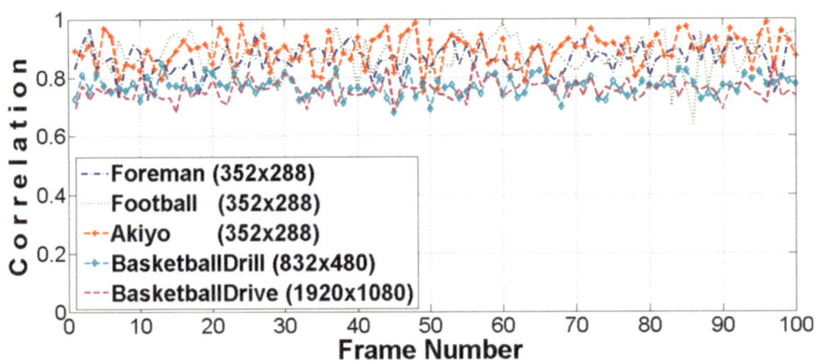

Figure 12.
Normalized correlations of the proposed watermarking process.

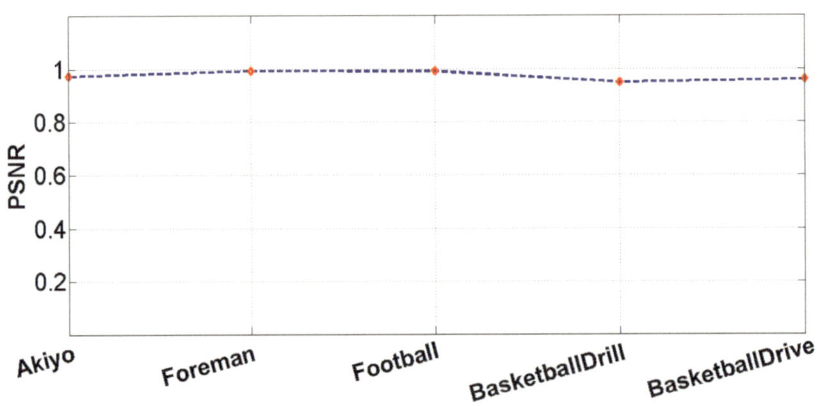

Figure 13.
Correlation values of the extracted watermarks using our detection process.

correlation values using our detection algorithm, while **Figure 14** shows the PSNRs of the reconstructed frames. **Figure 15** shows the PSNRs for the proposed watermarking method and the method of [20] when different numbers of frames of Football video were used. The method of [20] gives a maximum PSNR of 30 dB for the Football video, while our method gives an average of 42 dB; moreover, the performance in terms of NCs for the Football video stream using our method and the method in [20] was evaluated. The NC of method in [20] has an average value of 0.73, while our method gave a smooth performance with an average value of 0.99; this was shown as well in **Figure 13**.

For further investigation and evaluation of the robustness of our technique, the test videos will be subjected to some familiar attacks. These include additive noise, cropping, sharpening, rotating, frame averaging process, and HEVC compression. The attacks have the following characteristics:

- The additive noise will be Gaussian with a mean of 0 and a variance of 0.01.
- The salt-and-pepper noise has 1% noise density.
- Twenty-five percent of the even frames will be cropped.
- All the frames will be sharpened.
- The even frames will be rotated 1 degree counterclockwise.
- Ten random frames will be averaged with their respective successors.

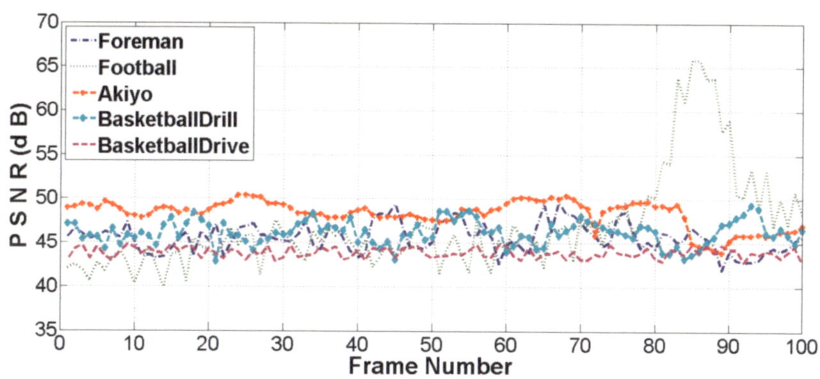

Figure 14.
PSNRs of the test videos when the proposed watermarking process is used.

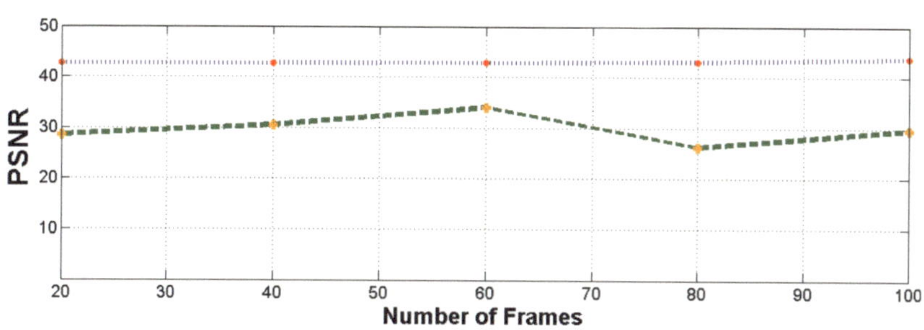

Figure 15.
PSNRs of the proposed method (the blue line) and method of [20] (the green line) when different numbers of frames of the video sequence Football are used.

The results for each attack for the different standard videos are shown in **Figure 16**.

To ensure the robustness of the algorithm, it was tested with the application of HEVC process. HEVC process with a quantization parameter (QP) value of 20 was applied to 100 frames of the test videos. Different compression ratios will result depending on each input video when this quantization factor is used. First, the watermarking process was applied to the test video frames and without applying the enhancement process. The PSNRs are shown in **Figure 17**, and the NCs at the watermarked frames are shown in **Figure 18**. A significant observation here is that lower values will result in the frames between 80 and 90 for the Football video, and

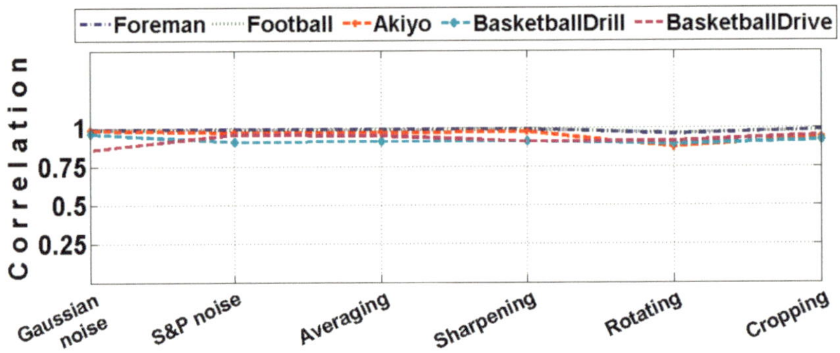

Figure 16.
Performance of the watermarking process under common aggressive attacks.

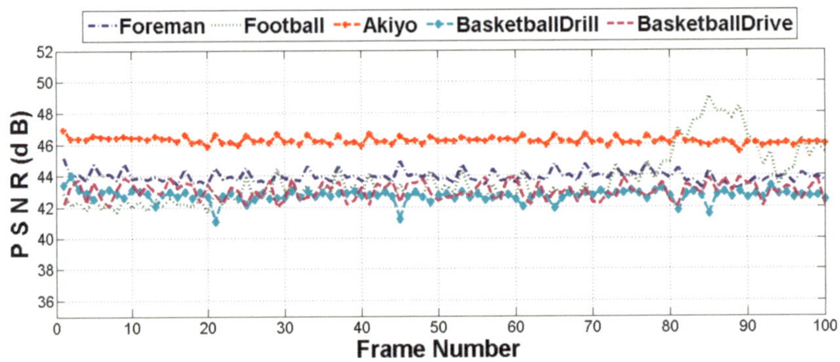

Figure 17.
PSNRs of the proposed watermarking method when HEVC process is applied.

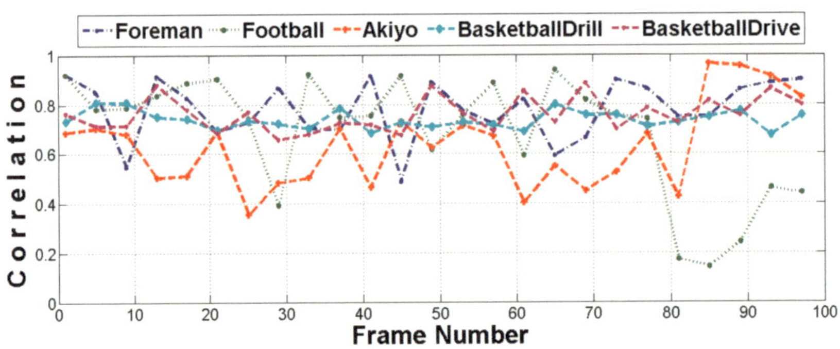

Figure 18.
Normalized correlations of the proposed watermarking method when HEVC process is applied.

this can be attributed to the blurring effects in these frames as a result of the fast panning of the camera. This, in turn, would result in lower energy in the DWT coefficients, and it is possible to overcome this phenomenon by our enhanced detection process as we will see later.

The correlations, moreover, were evaluated when the proposed detection process was applied. The watermarks were embedded randomly in multiple frames; then they were extracted and processed according to the method in Section 3. A number of frames being used for embedding process are 20%, 50%, and ultimately 100% of the 100 test frames of the proposed standard videos. **Figure 19** shows the performance under these circumstances. It can be shown that the detection process was enhanced dramatically when comparing with **Figure 18**. The system can perform well even with only 20% of the frames being watermarked. To evaluate our algorithm under different compression ratios, the standard test videos were watermarked and compressed using HEVC with different QPs: 15, 20, and 25. The value of 20 is a typical value for compression. **Figure 20** shows the performance of our system under these compression values. As QP becomes larger than 25, the video qualities go through noticeable degradation in terms of resolutions; in fact, the system performs well for QP values of 20 or less, and as QP values reach 25, the detection process starts to lose its efficiency for some videos. This is due to the aggressive quantization process of the discrete cosine transform (DCT) coefficients in the HEVC process.

Our selective denoising filter which was introduced in Section 4 was tested for the watermarked videos. The standard videos were watermarked according to the proposed embedding process; then they were subjected to salt-and-pepper noise

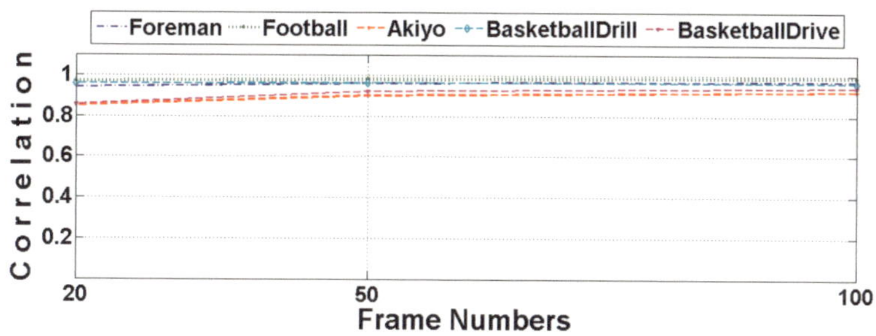

Figure 19.
Correlations vs. number of frames being watermarked when our detection method is used.

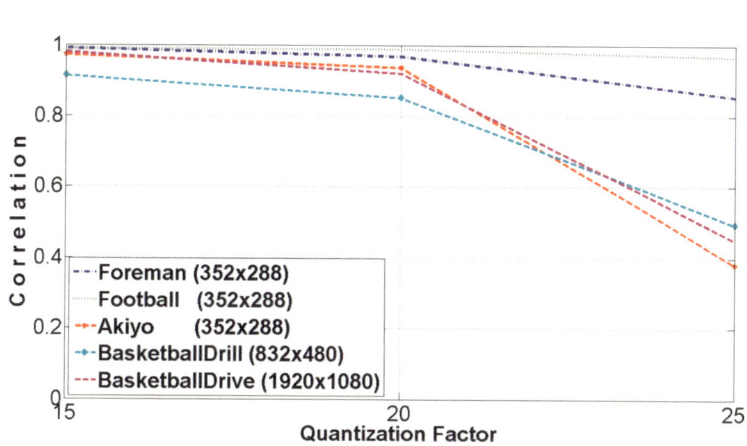

Figure 20.
Watermarking process performance under different quantization parameters.

with 1% density. At the receiving side, the videos were denoised using the selective filter; then the watermarks were extracted. **Figure 21** shows the normalized correlations and the PSNRs under these conditions. It is clear that the selective filtering scheme enhanced the visual appearance by eliminating the noise without significant effects on the efficiency of the watermarking process. Furthermore, **Figure 22** shows the results when 2D median filter is used. It can be seen that our selective denoising filter outperformed the 2D median filter in terms of the correlation

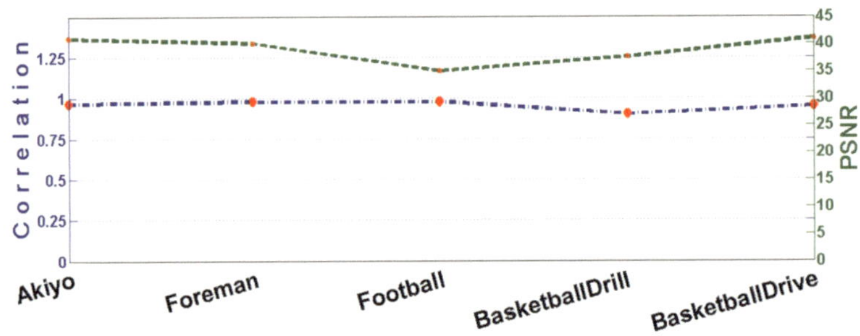

Figure 21.
The performance of the watermarking process with the use of the proposed selective filter where the blue line is the correlation and the green line is the PSNR.

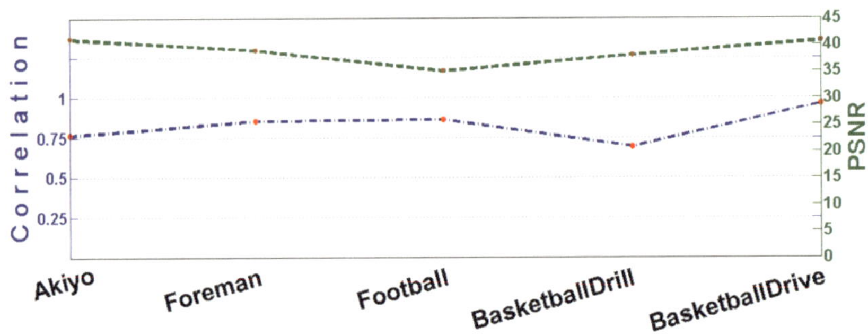

Figure 22.
The performance of the watermarking process with the use of a 2D median filter where the blue line is the correlation and the green line is the PSNR.

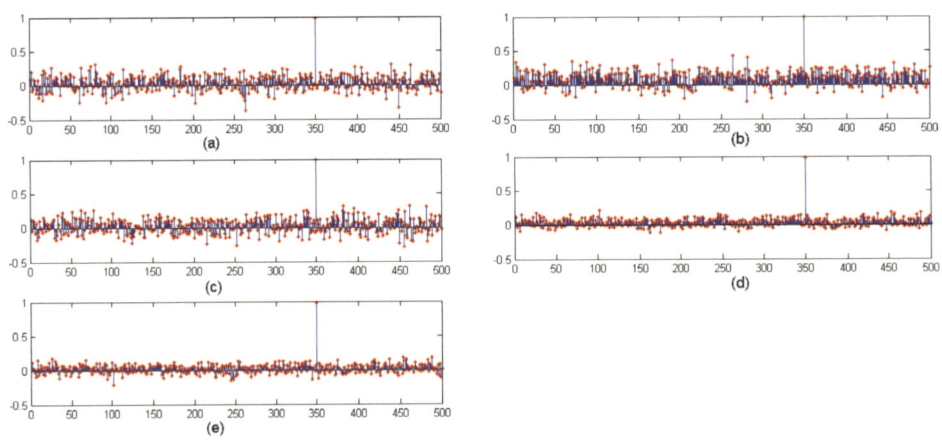

Figure 23.
The watermarking process response to false alarm test, the right watermark is the 350th with different videos: (a) Akiyo, (b) Foreman, (c) Football, (d) BasketballDrill, and (e) BasketballDrive.

values. The PSNR values are almost comparable, even though when assessing the quality of the images, the metric parameters are not the only factor that should be taken into consideration. The perceptual quality is another factor which was better with our own filter due to the selectivity process.

Since the security is an important aspect for our algorithm and any data hiding technique, false alarm attacks were studied for our test videos. That happens when no hiding process was done or a false watermark was hidden and still the system indicates the existence of our watermark. To do that, we generate 500 different random watermarks and hide them in the test videos according to our proposed algorithm, and the right watermark was one of them and it was set to be the 350th one. **Figure 23** shows the results. It can be seen that the response of our system was low to the false watermarks, and only the right watermark resulted in high response. This is an indication of good reliability of our system.

6. Conclusions and future work

This work proposes a DWT-based watermarking process using randomly generated orthonormal filter banks. An enhanced detection process was proposed to add to the robustness of the system. Moreover, a selective filtering process was developed to eliminate the noise. A good deal of the security of the system was achieved by the randomness in the filter banks, the pseudorandom sequence that was used to encode the watermark, and the regions of hidings. It was shown that the proposed technique performs well with and without HEVC. The compression ratio that was used is typical. Further investigation of the efficiency of the watermarking process under other aggressive attacks will be discussed and researched in future work. Moreover, an integration process of the data hiding process inside videos and the HEVC process will be studied and investigated.

Conflict of interest

The authors whose names are listed above certify that there are no conflicts of interests of any sort or nature between any of them and any institution or organization (public or private) that have special interest in the research that is the topic of this work.

Author details

Farhan Al-Enizi[1*] and Awad Al-Asmari[2]

1 PSAU University, Al-Kharj, Saudi Arabia

2 Shaqra University, Saudi Arabia

*Address all correspondence to: farhan414@gmail.com

References

[1] Langelaar GC, Setyawan I, Lagendijk RL. Watermarking digital image and video data. A state-of-the-art overview. IEEE Signal Processing Magazine. 2000; **17**(5)

[2] Sullivan GJ, Boyce JM, Chen Y, Ohm J-R, Segall CA, Vetro A. Standardized extensions of high efficiency video coding (HEVC). Selected Topics in Signal Processing, IEEE Journal of. 2013; 7(6):1001-1016

[3] Sullivan GJ, Ohm J-R, Han W-J, Wiegand T. Overview of the high efficiency video coding (hevc) standard. IEEE Transactions on Circuits and Systems for Video Technology. 2012; **22**(12):1649-1668

[4] Panyavaraporn J. Multiple video watermarking algorithm based on wavelet transform. In: 2013 13th International Symposium on Communications and Information Technologies (ISCIT), IEEE. 2013. pp. 397-401

[5] Meerwald P, Uhl A. Survey of wavelet-domain watermarking algorithms. In: Photonics West 2001-Electronic Imaging. International Society for Optics and Photonics; 2001. pp. 505-516

[6] Lee M-S. Image compression and watermarking by wavelet localization. International Journal of Computer Mathematics. 2003;**80**(4):401-412

[7] Kundur D, Hatzinakos D. Digital watermarking using multiresolution wavelet decomposition. In: Proceedings of the 1998 IEEE International Conference on Acoustics, Speech and Signal Processing; 1998; vol. 5; IEEE. 1998. pp. 2969-2972

[8] Guzmán VVH, Miyatake MN, Meana HMP. Analysis of a wavelet-based watermarking algorithm. In: 14th International Conference on Electronics, Communications and Computers, 2004. CONIELECOMP 2004; IEEE. 2004. pp. 283-287

[9] Wang S-H, Lin Y-P. Wavelet tree quantization for copyright protection watermarking. IEEE Transactions on Image Processing. 2004;**13**(2):154-165

[10] Bhattacharya S, Chattopadhyay T, Pal A. A survey on different video watermarking techniques and comparative analysis with reference to h. 264/avc. In: 2006 IEEE Tenth International Symposium on Consumer Electronics, 2006. ISCE'06. IEEE. 2006. pp. 1-6

[11] Martinez R, Reyes R, Cruz C, Nakano M, Perez H. A dwt-based video watermarking scheme resilient to mpeg-2 compression and collusion attacks. In: International Symposium on Information Theory and Its Applications 'ISITA 2008', IEEE. 2008. pp. 1-5

[12] Abdallah EE, Hamza AB, Bhattacharya P. Video watermarking using wavelet transform and tensor algebra. Signal, Image and Video Processing. 2010;**4**(2):233-245

[13] Vetterli M, Kovacevic J. Wavelets and Subband Coding; 1995

[14] Panyavaraporn J. Wavelet based video watermarking scheme for h. 264/avc. In: 2011 International Symposium on Intelligent Signal Processing and Communications Systems (ISPACS); IEEE. 2011. pp. 1-5

[15] Xu D, Wang R, Wang J. Video watermarking based on spatio-temporal jnd profile. In: International Workshop on Digital Watermarking; Springer. 2008. pp. 327-341

[16] Vinod P, Bora P. A new inter-frame collusion attack and a countermeasure.

In: International Workshop on Digital Watermarking; Springer. 2005. pp. 147-157

[17] Liu Y, Zhao J. A new video watermarking algorithm based on 1d dft and radon transform. Signal Processing. 2010;**90**(2):626-639

[18] Matsubara T, Moshnyaga VG, Hashimoto K. A low-complexity and low power median filter design. In: 2010 International Symposium on Intelligent Signal Processing and Communication Systems. 2010

[19] Deivalakshmi S, Sarath S, Palanisamy P. Detection and removal of salt and pepper noise in images by improved median filter. In: Recent Advances in Intelligent Computational Systems (RAICS), 2011 IEEE; IEEE. 2011. pp. 363-368

[20] Sundararajan M, Yamuna G. Dwt based scheme for video watermarking. In: 2013 International Conference on Communications and Signal Processing (ICCSP); IEEE. 2013. pp. 460-464

Chapter 5

Wavelet Transform Analysis to Applications in Electric Power Systems

Mario Orlando Oliveira, José Horacio Reversat and Lucas Alberto Reynoso

Abstract

The wavelet transform has received great importance in the last years on the power system analysis because the multi-resolution analysis presents proprieties good for the transient signal analysis. This chapter presents a review on main application of wavelet transform in electric power systems. The study areas have been classified as power system protection, power quality disturbances, power system transient, partial discharge, load forecasting, faults detection, and power system measurement. The areas in which more works have been developed are the power quality and protections field, where both cover 51% of the articles analyzed.

Keywords: electric power systems, wavelet transform, signal processing, transient events

1. Introduction

Electromagnetics transients in electric power systems (EPS) are generally caused by lightning discharges and/or certain operating conditions, such as faults in equipment and transmission lines, switching of electric power system devices, voltage sags, capacitor switching, and transmission line energization and de-energization. Faulted EPS signals are associated with fast electromagnetic transients and are typically nonperiodic with high-frequency oscillations. These characteristics present a problem for traditional Fourier analysis because it assumes a periodic signal and a wide-band signal that require denser sampling and longer time periods to maintain good resolution in low frequencies. Wavelet transform (WT), on the other hand, is a powerful tool in the analysis of transient phenomena in power systems. It has the ability to extract information from the transient signals simultaneously in both time and frequency domains and has replaced the Fourier analysis in many applications. This ability to tailor the frequency resolution can greatly facilitate the detection of signal features that may be useful in characterizing the transient cause or the state of the postdisturbance electrical system.

On the other hand, the waveforms associated with fast electromagnetic transients are typically nonperiodic and contain both high frequency oscillations and localized superimposed impulses on power frequency and its harmonics. These characteristics present problems for traditional Fourier analysis because the latter assumes a periodic signal that needs longer time periods to maintain good resolution in the low frequency. In this sense, WT has received great attention in power community in the last years because they are better suited for the analysis of certain types of transient waveforms than the other transform approaches.

Many books and papers have been written that explain WT of signals and can be read for further understanding of the basics of wavelet theory. The first recorded mention of what we now call a "wavelet" seems to be in 1909, in a thesis by A. Haar. The concept of wavelets in its present theoretical form was first proposed by J. Morlet, a Geophysicist, and the team at the Marseille Theoretical Physics Center working under A. Grossmann, a theoretical physicist, in France. They provided a way of thinking for wavelets based on physical intuition. In other words, the transform of a signal does not change the information content presented in the signal [1].

Thus, in the first part, this chapter presents an overview of the main characteristic of wavelet transform for the transient signal analysis and the application on electric power system. The property of multiresolution in time and frequency provided by wavelets allows accurate time location of transient components while simultaneously retaining information about the fundamental frequency and its low-order harmonics. This property of the wavelet transform facilitates the detection of physically relevant features in transient signal to characterize the source of the transient or the state of the postdisturbance system.

Initially, we will discuss the performance, advantages, and limitations of the WT in electric power system application, where the basic wavelet theory is presented. Additionally, the main publications carried out in this field will be analyzed and classified by the next areas: power system protection, power quality disturbances, power system transient, partial discharge, load forecasting, faults detection, and power system measurement. Finally, a comprehensive analysis related to the advantages and disadvantages of the WT in relation to other tools is performed.

2. Wavelet transform (WT)

The wavelet transform theory is based on analysis of signal using varying scales in the time domain and frequency. Formalization was carried out in the 1980s, based on the generalization of familiar concepts. The wavelet term was introduced by French geophysicist Jean Morlet. The seismic data analyzed by Morlet exhibit frequency component that changed rapidly over time, for which the Fourier Transform (FT) is not appropriate as an analysis tool. Thus, with the help of theoretical physicist Croatian Alex Grossmann, Morlet introduced a new transform which allows the location of high-frequency events with a better temporal resolution [2].

Faulted EPS signals are associated with fast electromagnetic transients and are typically nonperiodic and with high-frequency oscillations. This characteristic presents a problem for traditional Fourier analysis because it assumes a periodic signal and a wide-band signal that require more dense sampling and longer time periods to maintain good resolution in the low frequencies [3]. The WT is a powerful tool in the analysis of transient phenomena in power system. It has the ability to extract information from the transient signals simultaneously in both time and frequency domains and has replaced the Fourier analysis in many applications [4].

2.1 Continuous wavelet transform (CWT)

The short-time Fourier transform (STFT) of the continuous signal $x(t)$ can be seen as the Fourier Transform (FT) of the signal with windowed $x(t).g(t-\tau)$ or also as a signal decomposition $x(t)$ into basis functions $g(t-\tau).e^{-jwt}$. The functions based term refers to a complete set of functions that, when combined on the sum with specific weight can be used to then construct a certain sign [5].

In the FT case, the base functions are complex sinusoid e^{-jwt} with a windows centered on the τ time. The WT is described in terms of its basic functions, called wavelet or mother wavelet, and variable frequency w is replaced by an ever-escalating variable factor a (which represents the swelling) and, generally, to variable displacement in time τ, is represented by b.

The main characteristic of the WT is that it uses a variable window to scan the frequency spectrum, increasing the temporal resolution of the analysis. The wavelets are represented by:

$$\psi_{a,b}(t) = \frac{1}{\sqrt{a}} \cdot \psi\left(\frac{t-b}{a}\right) \tag{1}$$

In Eq. (1), the constant $1/\sqrt{a}$ is used to normalize the energy and ensure that the energy of $\psi_{a,b}(t)$ is independent of the dilation level [6]. The wavelet is derived from operations such as dilating and translating the mother wavelet, ψ, which must satisfy the admissibility criterion given by [7]:

$$C_\psi = \int_{-\infty}^{+\infty} \frac{\left|\widehat{\psi}\ (y)\right|^2}{|y|} dy \leq \infty \tag{2}$$

where $\widehat{\psi}\ (y)$ is the FT of the $\psi\ (t)$. This means that if $\widehat{\psi}$ is a continuous function, then C_ψ is finite only if$\psi\ (0) = 0$, i.e., [7]:

$$\int_{-\infty}^{+\infty} \psi(t)\, dt = 0 \tag{3}$$

Thus, it is evident that WT has a zero rating property that increases the degrees of freedom, allowing the introduction of the dilation parameter of the window [8].

The continuous wavelet transform (CWT) of the continuous signal $x(t)$ is defined as:

$$(CWT)(a,b) = \int_{-\infty}^{+\infty} x(t) \cdot \psi_{a.b}(t)\, dt = \frac{1}{\sqrt{a}} \int_{-\infty}^{+\infty} x(t) \cdot \psi\left(\frac{t-b}{a}\right) dt \tag{4}$$

where the scale factor a and the translation factor b are continuous variables.

The WT coefficient is an expansion and a particular shift represents how well the original signal $x(t)$ corresponds to the translated and dilated mother wavelet. Thus, the coefficient group of CWT(a,b) associated with a particular signal is the wavelet representation of the original signal $x(t)$ in relation to the mother wavelet [9].

2.2 Discrete wavelet transform (DWT)

2.2.1 DWT definition

The redundancy of information and the enormous computational effort to calculate all possible translations and scales of CWT restricts its use. An alternative to this analysis is the discretization of the scale and translation factors, leading to the

DWT. There are several ways to introduce the concept of DWT, the main are the decomposition bands and the decomposition pyramid (or Multi-Resolution Analysis -MRA), developed in the late 1970s [10]. The DWT of the continuous signal $x(t)$ is given by:

$$(DWT)(m,p) = \int_{-\infty}^{+\infty} x(t) \cdot \psi_{m,p} dt \tag{5}$$

where $\psi_{m,p}$ form bases of wavelet functions, created from a translated and dilated of the mother wavelet using the dilation m and translation p parameters, respectively.

Thus, $\psi_{m,p}$ is defined as:

$$\psi_{m,p} = \frac{1}{\sqrt{a_0^m}} \psi \left(\frac{t - pb_0 a_0^m}{a_0^m} \right) \tag{6}$$

The DWT of a discrete signal $x[n]$ is derived from CWT and defined as [9]:

$$(DWT)(m,k) = \frac{1}{\sqrt{a}} \sum_n x[n] \cdot g \left(\frac{k - nb_0 a_0^m}{a_0^m} \right) \tag{7}$$

where $g(*)$ is the mother wavelets and $x[n]$ is the discretized signal function.

The mother wavelets may be dilated and translated discretely by selecting the scaling and translation parameters $a = a_0^{\mathrm{m}}$ and $b = nb_0 a_0^m$ respectively (with fixed constants $a_0>1$, $b_0>1$, m and n belonging the set of positive integers).

2.2.2 Multi-resolution analysis (MRA)

The problems of temporal resolution and frequency found in the analysis of signals with the STFT (best resolution in time at the expense of a lower resolution in frequency and vice-versa) can be reduced through a multi-resolution analysis (MRA) provided by WT. The temporal resolutions, Δt, and frequency, Δf, indicate the precision time and frequency in the analysis of the signal. Both parameters vary in terms of time and frequency, respectively, in signal analysis using WT. In the STFT, a higher temporal resolution could be achieved at the expense of frequency resolution. Intuitively, when the analysis is done from the point of view of filters series, the temporal resolution should increase increasing the center frequency of the filters bank. Thus, tax is that Δf is proportional to f, i.e.,

$$\frac{\Delta f}{f} = c \tag{8}$$

where c is constant.

The main difference between DWT and STFT is the time-scaling parameter. The result is geometric scaling, i.e., 1, $1/a$, $1/a^2$, ...; and translation by 0, n, $2n$, and so on. This scaling gives the DWT logarithmic frequency coverage in contrast to the uniform frequency coverage of the STFT, as compared in **Figure 1**.

The CWT follows exactly these concepts and adds the simplification of the scale, where all the impulse responses of the filter bank are defined as dilated versions of a mother wavelet [10]. The CWT is a correlation between a wavelet at different scales and the signal with the scale (or the frequency) being used as a measure of similarity. The CWT is computed by changing the scale of the analysis window,

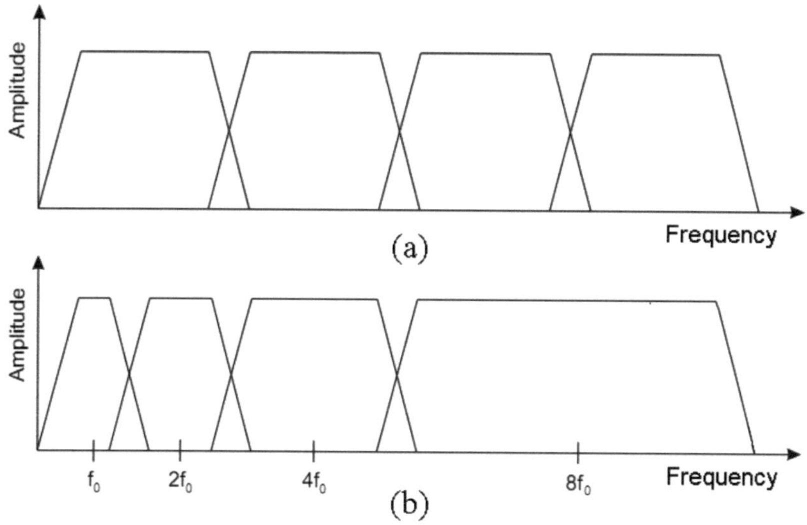

Figure 1.
Comparison of (a) the STFT uniform frequency coverage to (b) the logarithmic coverage of the DWT.

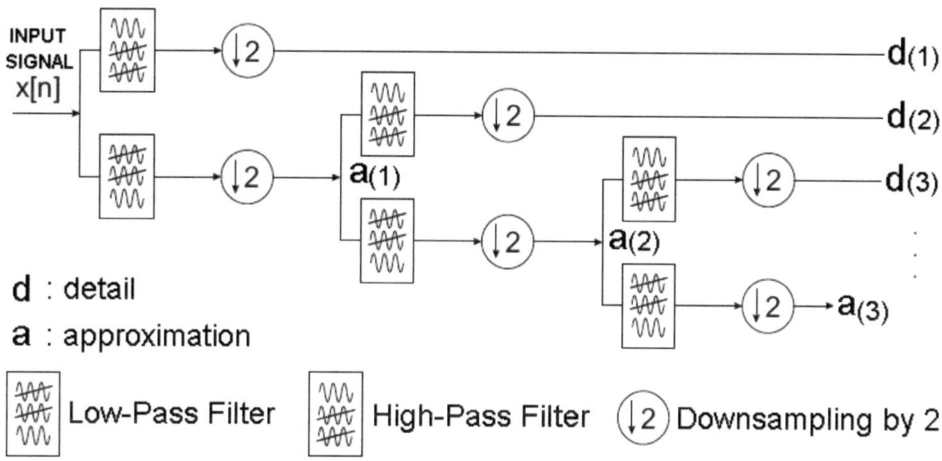

Figure 2.
DWT filter bank framework.

shifting the window in time, multiplying by the signal, and integrating over all times. In the discrete case, filters of different cut-off frequencies are used to analyze the signal at different scales. The signal is passed through a series of high-pass filters to analyze the high frequencies, and it is passed through a series of low-pass filters to analyze the low frequencies. Thus, the DWT can be implemented by multistage filter bank named MRA [11], as illustrated on **Figure 2**. The Mallat algorithm consists of series of high-pass and the low-pass filters that decompose the original signal $x[n]$ into approximation a(n) and detail d(n) coefficient, each one corresponding to a frequency bandwidth.

The resolution of the signal, which is a measure of the amount of detail information in the signal, is changed by the filtering operations, and the scale is changed by up-sampling and down-sampling (sub-sampling) operations. Sub-sampling a signal corresponds to reducing the sampling rate or removing some of the samples of the signal. For other hand, up-sampling a signal corresponds to increasing the sampling rate of a signal by adding new samples to the signal.

The procedure starts with passing this signal $x[n]$ through a half band digital low-pass filter with impulse response $h[n]$. Filtering a signal corresponds to the mathematical operation of convolution of the signal with the impulse response of the filter. The convolution operation in discrete time is defined as follows [2]:

$$x[n] * h[n] = \sum_{k=-\infty}^{\infty} x[k] \cdot h[n-k] \tag{9}$$

A half band low-pass filter removes all frequencies that are above half of the highest frequency in the signal. For example, if a signal has a maximum of 1000 Hz component, then half band low-pass filtering removes all the frequencies above 500 Hz. However, it should always be remembered that the unit of frequency for discrete time signals is radians.

After passing the signal through a half band low-pass filter, half of the samples can be eliminated according to the Nyquist's rule. Simply discarding every other sample will subsample the signal by two, and the signal will then have half the number of points. The scale of the signal is now doubled. Note that the low-pass filtering removes the high frequency information but leaves the scale unchanged. Only the sub-sampling process changes the scale. Resolution, on the other hand, is related to the amount of information in the signal, and therefore, it is affected by the filtering operations. Half band low-pass filtering removes half of the frequencies, which can be interpreted as losing half of the information. Therefore, the resolution is halved after the filtering operation. Note, however, the sub-sampling operation after filtering does not affect the resolution, since removing half of the spectral components from the signal makes half the number of samples redundant anyway. Half of the samples can be discarded without any loss of information.

This procedure can mathematically be expressed as [2]:

$$y[n] = \sum_{k=-\infty}^{\infty} h[k] \cdot x[n-k] \tag{10}$$

The decomposition of the signal into different frequency bands is simply obtained by successive high-pass and low-pass filtering of the time domain signal. The original signal $x[n]$ is first passed through a half band high-pass filter $g[n]$ and a low-pass filter $h[n]$. After the filtering, half of the samples can be eliminated according to the Nyquist's rule, since the signal now has a highest frequency of $p/2$ radians instead of p. The signal can therefore be sub-sampled by 2, simply by discarding every other sample. This constitutes one level of decomposition and can mathematically be expressed as follows [2]:

$$y_{high}[k] = \sum_{n} x[n] \cdot g[2k-n] \tag{11}$$

$$y_{low}[k] = \sum_{n} x[n] \cdot h[2k-n] \tag{12}$$

where $y_{high}[k]$ and $y_{low}[k]$ are the outputs of the high-pass and low-pass filters, respectively, after sub-sampling by 2.

3. Wavelets theory advantage

In [12], an application of WT and its advantages compared to Fourier transform is presented. One of the main advantages of wavelets is that they offer a simultaneous localization in time and frequency domain. The second main advantage of wavelets is that, using fast wavelet transform, it is computationally very fast. Wavelets have the great advantage of being able to separate the fine details in a signal. Very small wavelets can be used to isolate very fine details in a signal, while very large wavelets can identify coarse details. A wavelet transform can be used to decompose a signal into component wavelets. In wavelet theory, it is often possible to obtain a good approximation of the given function f by using only a few coefficients, which is a great achievement when compared to Fourier transform. Wavelet theory is capable of revealing aspects of data that other signal analysis techniques miss like trends, breakdown points, and discontinuities in higher derivatives and self-similarity. It can often compress or de-noise a signal without appreciable degradation [12].

The Fourier transform shows up in a remarkable number of areas outside classic signal processing. Even taking this into account, we think that it is safe to say that the mathematics of wavelets is much larger than that of the Fourier transform. In fact, the mathematics of wavelets encompasses the Fourier transform. The size of wavelet theory is matched by the size of the application area. Initial wavelet applications involved signal processing and filtering. However, wavelets have been applied in many other areas including nonlinear regression and compression. An offshoot of wavelet compression allows the amount of determinism in a time series to be estimated [12].

4. Wavelets transform application in electric power system

Refs. [1, 13, 14] conducted studies related to this chapter. These authors also present a literature review on the application of WT in power electrical systems.

By means of the bibliographic review, it is possible to highlight certain topics of interest for researchers:

- Power quality
- Partial discharges
- Transient in electrical systems
- Power system protection
- Load forecasting
- Power system measurement

Figure 3 shows the percentage of publications in each area. The areas in which more works have been developed are the power quality and protection field. The next section presents a general description of wavelet application in the selected areas of power systems. There are more works in these areas; however, no details will be entered due to space issues and that the approach to the topic used WT is similar.

4.1 Partial discharges

Partial discharges are difficult to detect because of their short duration, high frequency, and low amplitude. However, the use of WT can not only detect them

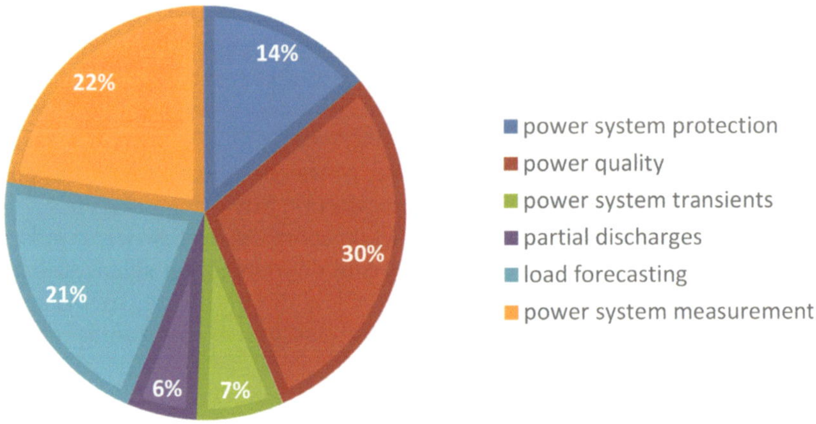

Figure 3.
Wavelet publication percentage in different power system areas.

but also isolate them by frequency bands for their study, performing a multi-resolution analysis. An important study is the filtering of noise in PD signals. In this sense, [15, 16] propose the recognition and categorization of PDs, for the study of transformer failures.

In [15], an adaptive filter is used to obtain PD without noise or interference for online studies. In [16], it is posed by means of a sophisticated equipment to obtain the currents and through WT to filter the PD and to be able to carry out its later study. If we follow the line of online monitoring, it can be seen in [17] that using a high-frequency transducer, the PD can be extracted to obtain analysis in transformers in service. We talked about how to get them and how to remove the noise in the signals, we just need to locate them. Ref. [18] proposes a PD injection method in preestablished nodes to obtain the plant model to measure at separate points.

4.2 Transient in electrical system

This section could be assembled considering different places of application of the WT, from generation to service transformers. In the generation, [19] shows us a study in the transients of the generation turbines where they perform an exhaustive analysis using a wavelet neural network to obtain the output values to adjust the turbine and obtain a correct operation.

Works [20–22] propose the use of WT for applications in high voltage lines (EHV) and the power system. In [20], WT used for the study of the electric field in the conductors can obtain the waveform of the current and voltage that facilitates the study of faults with which a filter for the study of harmonics can be designed. In [21], using the entropy energy, it is possible to obtain parameters of systems that change with certain sensitivity and design a protection that can reduce the capacitive effects in the bars and in the high-frequency traps. Ref. [22] proposes to use several scales of different frequencies to decompose the harmonics for the detection, localization, and segmentation of them. With this, we can estimate the energy and overvoltage caused and discern between impulsive and oscillatory transients. Ref. [23] presents an analysis on which is the best wavelet mother for the measurement of harmonics in electrical systems. Ref. [24] proposes an empirical wavelet transform for harmonic detector under dynamism conditions of the system. Ref. [25] presents a method for detecting and classifying faults in transmission lines by combining DWT and neural networks.

Finally, the characteristic impedance of a transformer can be obtained through its transient response [26]. With this, we can easily make a verification of the state of the transformer using the WT coefficients.

4.3 Power system protection

In the search for shorter downtimes and the maximization of the life time of the energy system equipment, it repeats itself to increasingly finer methods for analysis. In the recent years, we have begun to analyze the protections of voltage transformers (TT) and current transformers (CT) by WT [27]. This was done previously in the use of the Fourier Transform (FFT) and the current protections of the internal fault current. Refs. [28–31] study the differentiation of these currents by theoretical methods. This means that the test in the field is not done but the results of the simulations give positive results, an exception of [31] where the methods are applied to a test TT. For the development of differential protections, [30, 32–34] analyze the transients and extract the predominant signal from the internal fault; the method has been tested with different faults and an efficient algorithm for online analysis has been obtained. In [33–41], WT is used for fault diagnostic and transient event identification in transformer operation.

There has been a great effort in different works to evaluate the different mother wavelet. Thus, [33, 34] present a comprehensive analysis involving an important number of wavelets to prove the efficiency in power transformer protection.

Another important part is between disruptive currents and fault currents in the earth, where the results were analyzed with computer programs [42, 43].

The bus zone protection scheme is considered for the detection of transient current in [44]. In the electric machine area, [45] proposed the use of the WT for fault diagnostic using current signal. The transmission line fault detection with WT in the presence of wind power generation is studied in [46, 47] approach, a study on fault detection in compensated transmission line (TCSC technology). In [48], wavelet analysis is used for fault detection in hybrid energy source.

4.4 Power quality

With the increase of the use of electronic equipment, it has become a necessity to study the quality of energy. The loads, having a stationary or non-stationary state, are no longer easily analyzed, so we resort to mathematical tools capable of classifying and characterizing these states. In this network, transient disturbance occurring in the network can be classified as voltage drops, voltage increases, momentary interruptions, harmonics, and transients, among others.

The main topics of study are the harmonics; these are present in the waves and are common to see in the electrical networks, and so the first step would be to identify them and isolate their behavior in order to eliminate these disturbances. Ref. [49–51], using WPT, perform a method in which the fundamental harmonic and higher order harmonics can be extracted, which is a method applicable to energy networks for monitoring them. For the same application [52], they have formulated using CWT, an algorithm used to identify the variations of frequency, harmonics, sub harmonics, and inter harmonics.

A practical application of the CWT is the real-time monitoring of voltage and current signals for the rapid detection and elimination of transient events that may worsen the quality of the electric service. Previously, the first subject of study in terms of energy quality was established, but this is only a small part of the faults found in the energy system. Faults must be identified in order to perform an efficient analysis and be able to act on them as soon as they occur.

In the papers presented in [42, 53–61], their main theme coincides and is to classify the faults that can be obtained from the analysis of the voltage wave through WT. Broadly speaking, it can be highlighted that they obtain the range of frequency, amplitude, time of action, waveform, etc. Although their methods differ in the application of the WT, the essence of the study is the same. One step beyond performing fault classification studies is to achieve a practical application for the in-service power network. In [62], it implements a DSP for the continuous study of energy. Ref. [63] uses an FPGA for continuous analysis of system disturbances. In [42], through a wavelet neural network and a self-organizing arrangement system, it obtains automatic equipment for the detection and suppression of multiple faults in the network.

Other areas of power quality are studied using the WT characteristic. In [64, 65], the power quality events are characterized and classified using wavelet transformation. The power quality disturbance detection in grid-connected wind energy system is development with wavelet and S-transform. In [66] is presented a comparative analysis of power quality event using wavelet for real time implementation and [67] tested measurement system to electric energy quality together digital signal processor.

The voltage variation is a usual problem in electric system which affects the quality power. Ref. [68] addresses this problem from distributed energy resources.

4.5 Load forecasting

This analysis is the key to a correct distribution of energy in the electricity grid. Basically, a prediction is made of how the loads will behave in short- or medium-term horizons. In this sense, the behavior of the load can refer to disconnections of large equipment, circuits that have transients in start, devices with low power factor, etc. This analysis can be done through WT and some complement to make the prediction.

We have cases like [69] that use a linear correlation for the load forecast. This is a more conventional method compared to the other two. In recent years, there is much talk of NN and machine learning to such a point that [70–79] use machine learning to train a WT-based neural network. The load forecasting in the short-term used WT theory is presented in [80, 81]. Thus, it can be observed that this application is widely studied.

From another point of view, we have a prediction system based on fuzzy logic in [82], and finally, in the current year, a neural network with fuzzy logic based on WT is launched for the short-term study, which may be the foot to what is coming to the future.

4.6 Power system measurements

Mainly, the WT is used in power systems and RMS measurements, both voltage and current. As for the measurement systems of active, reactive, or apparent power, a decomposition and classification of those waves present in the fundamental are achieved. In [83], it analyzes how the variations of the loads influence the tension, as much as it falls as in descents or in blinking. Then in [23], it approximates the RMS value of the harmonics present in the voltage and current waves by means of statistics; with this purpose, the improvement of harmonic detection and measurement systems is sought. In [84], a method for determination and correction of measurement anomaly based in WT is proposed.

5. Conclusion

Wavelet transform is a powerful signal processing tool that transforms a time-domain waveform into time-frequency domain and estimates the signal in the time and frequency domains simultaneously. So, it is mostly used in electric power systems analysis.

This chapter carries out an approach on the WT application in EPS in order to facilitate the search for information in this area. Therefore, a classification of the different fields of EPS applications was made. A summary description for each WT application area is presented with the main objective of showing the applications of this tool in the resolution of typical problems of the energy system.

The works analyzed show that the Daubechies family was used in most of the applications in power systems analysis, especially in protection area. However, the type of mother wavelet and the decomposition level number may be changed and therefore may not be generalized to all the cases.

The use of WT together with artificial intelligence tools (neural networks, fuzzy logic, genetic algorithms, etc.) was presented as a promising methodology to diagnose faults in electrical systems.

Most of the works analyzed results of computational simulations. It is expected that in the short term, the functionality of WT it is will comprobed in real situations.

Author details

Mario Orlando Oliveira*, José Horacio Reversat and Lucas Alberto Reynoso
Research and Development Laboratory in Electric Energy – LIDEE, Faculty of Engineering, National University of Misiones – UNaM, Oberá, Misiones, Argentina

*Address all correspondence to: oliveira@fio.unam.edu.ar

References

[1] Shariatinasab R, Akbari M, Rahmani B. Application of wavelet analysis in power systems. In: Baleanu D, editor. Advance in Wavelet Theory and their Application in Engineering, Physics and Technologic. Rijeka: InTech; 2012. ISBN 978-953-51-0494-0. pp. 221-244. DOI: 10.5772/2668

[2] Polikar R. The Story of Wavelets, Physics and Modern Topics in Mechanical and Electrical Engineering, World Scientific and Eng. USA: Society Press; 1999. pp. 192-197. Retrieved from: http://www.public.iastate.edu/~rpolikar

[3] Robertson DC, Camps OI, Mayer JS, Gish WB. Wavelets and electromagnetic power system transient. IEEE Transactions on Power Delivery. 1996, 1996;**11**(2):1050-1058. ISSN 0885-8977

[4] Phadke AG, Thorp JS. Computer Relaying for Power System. 2nd ed. Baldock, England: Research Studies Press Ltd. ISBN 978-0-470-05713-1; 2009

[5] Bentley PM, McDonnell JTE. Wavelets transform: An introduction. IEE Electronic & Communication Engineering Journal. 1994, 1994;**6**(4): 175-186. ISSN 0954-0695

[6] Simpson DM. An introduction to the discrete orthogonal wavelet transform. Revista Brasiliera de Engenharia. 1993, 1993;**9**(1):57-81

[7] Daubechies I. The wavelet transform, time-frequency localization and signal analysis. IEEE Transactions on Information Theory. 1990, 1990;**36**(5): 961-1005. ISSN 0018-9448

[8] Sarkar TK, Su C. A tutorial on wavelets from an electrical engineering perspective, part 2: The continuous case. IEEE Antennas and Propagation Magazine. 1998, 1998;**40**(6):36-49. ISSN 1045-9243

[9] Aggarwal R, Kim CH. Wavelet transform in power system : Part 1 – general introduction to the wavelet transform. Power Engineering Journal. 2000, 2007;**14**(2):81-87

[10] Rioul O, Vetterli M. Wavelets and signal processing. IEEE Signal Processing Magazine. 1991, 1994;**8**(4): 14-38. ISSN 1053-5888

[11] Mallat S. A Wavelet Tour of Signal Processing. 2nd ed. California, USA: Academic Press. ISBN 0-12-466606-X; 1999

[12] Sifuzzaman M, Islam MR, Ali MZ. Application of wavelet transform and its advantages compared to Fauirier transform. Journal of Physical Sciences. 2009;**13**:121-134 ISSN: 0972-8791

[13] Fernandez RC, Díaz Rojas HN. An overview of wavelets transform application in power system. In: Proceeding of 14th Power System Computation Conference (14th PSCC). Sevilla, España; 2002. Paper 6

[14] Akorede MF, Hizam H. Wavelet transform : Practical applications in power systems. Journal of Electrical Engineering and Technology. 2009; **4**(2):168-174. DOI: 10.5370/JEET.2009.4.2.168

[15] Zhao L, Sun Y. Noise elimination in partial discharge on-line monitoring for transformer based on wavelet transform. In: 2010 International Conference on E-Product E-Service and E-Entertainment, Henan, China, 7–9 Nov. 2010. pp. 1-4

[16] Mirzaei HR, Akbari A, Allahbakhshi M, Kharezi M. An experimental review on application of wavelet transform on partial discharge evaluation of power

transformers. In: 2010 10th IEEE International Conference on Solid Dielectrics, Potsdam, Germany, 4–9 July. 2010. pp. 1-4

[17] Seo J, Ma H, Saha TK. Differential partial discharge extraction technique for online power transformer insulation assessment. In: 2014 IEEE PES General Meeting | Conference & Exposition, National Harbor, MD, USA. 2014. pp. 1-5

[18] Sarikhani A, Darabi A, Banejad M. Partial discharge locating in power transformer using wavelet transform of bushing and grounded neutral currents. In: 2007 Australasian Universities Power Engineering Conference, Perth, WA, Australia. 2007. pp. 1-6

[19] Shanlin K, Yuzhe K, Huanzhen Z. Application of wavelet transform for fault pattern recognition and analysis of power system generator. In: 2009 Chinese Control and Decision Conference, Guilin, China. 2009. pp. 3908-3911

[20] Jaipradidtham C, Pasomkusolsil S. Harmonic analysis of electromagnetic transients in 500 kV single circuit transmission system using discrete wavelet transform. In: 2008 Joint International Conference on Power System Technology and IEEE Power India Conference, New Delhi, India. 2008. pp. 1-4

[21] Yang M, Yang Y. A study of transient-based protection using wavelet energy entropy for power system EHV transmission line. In: 2010 International Conference on Wavelet Analysis and Pattern Recognition, Qingdao, China. 2010. pp. 283-288

[22] Tan RHG, Ramachandaramurthy VK. Power system transient analysis using scale selection wavelet transform. In: TENCON 2009–2009 IEEE Region 10 Conference, Singapore, Singapore. 2009. pp. 1-6

[23] Kashyap S, Singh AK. Most suitable mother wavelet for measurement of power system harmonics using DWT in view of IEEE standard 1459-2000. In: 2008 13th International Conference on Harmonics and Quality of Power, Wollongong, NSW, USA. 2008. pp. 1-6

[24] Jain NL, Priyanka R, Keerthy P, Maya P, Babu P. Empirical wavelet transform for harmonic detection under dynamic condition. In: 2017 International Conference on Circuit , Power and Computing Technologies (ICCPCT), Kollam, India. 2017. pp. 1-5

[25] Ray P, Mishra DP, Dey K, Mishra P. Fault detection and classification of a transmission line using Discrete Wavelet Transform & Artificial Neural Network. In: 2017 International Conference on Information Technology (ICIT), Bhubaneswar, India. 2017. pp. 178-183

[26] Gomez-Luna E, Silva D, Aponte G, Pleite JG, Hinestroza D. Obtaining the electrical impedance using wavelet transform from the time response. IEEE Transactions on Power Delivery. 2013; **28**(2):1242-1244

[27] Rahmati A. An adaptive differential relay for CT saturation based on wavelet transform. In: IEEE PES T&D 2010, New Orleans, LA, USA. 2010. pp. 1-6

[28] Nagaopitakkul A, Kunakom A, Ngamroo I. Discrimination between external short circuits and internal faults in transformer windings using discrete wavelet transforms, Fourtieth IAS annual meeting. In: Conference Record of the 2005 Industry Applications Conference, 2005., Kowloon, Hong Kong, China. Vol. 1. 2005. pp. 448-452

[29] Yang L, Jingdong N. A wavelet transform based discrimination between internal faults and inrush currents in power transformers. In: 2011 International Conference on Electric

Information and Control Engineering, Wuhan, China. 2011. pp. 1127-1129

[30] Oliveira MO, Salim RH, Bretas AS. Differential protection of three-phase transformers using wavelet transforms. In: 2008 IEEE/PES Transmission and Distribution Conference and Exposition: Latin America, Bogota, Colombia. 2008. pp. 1-6

[31] Abbas MF, Zhiyuan L, Zhiguo H, Guanjun Z. Inrush current discrimination in power transformer differential protection using wavelet packet transform based technique. In: 2016 IEEE PES Asia-Pacific Power and Energy Engineering Conference (APPEEC), Xi'an. 2016. pp. 25, 944-28, 948

[32] Oliveira MO, Bretas AS. Application of discrete wavelet transform for differential protection of power transformer. In: Olkkonen H, editor. Discrete Wavelet Transform – Biomedical Applicattion. Rijeka: InTech; 2011. ISBN 978-953-307-354-6. pp. 349-366. DOI: 10.5772/1818

[33] Oliveira MO, Bretas AS, Ferreyra GD. Adaptative differential protection of three-phase power transformer based on transient signal analysis. Electric Power and Energy Systems. 2014;**57**: 366-374

[34] Ozgonenel O. Protection of power transformers by using continuous wavelet transform, MELECON 2006. In: 2006 IEEE Mediterranean Electrotechnical Conference, Malaga, Espain. 2006. pp. 1106-1109

[35] Abed NY, Mohammed OA. Modeling and characterization of transformers internal faults using finite element and discrete wavelet transforms. IEEE Transactions on Magnetics. 2007;**43**(4):1425-1428

[36] Suribabu, Ram S. Wavelet transform and ANN base differential protection for power transformer. In: 2016 International Conference on Signal Processing, Communication, Power and Embedded System (SCOPES), Paralakhemundi India. 2016. pp. 1882-1887

[37] Klomjit J, Ngaopitakkul A. Behaviour of interturn fault in power transformer winding using high frequency components of discrete wavelet transform. In: 2012 15th International Conference on Electrical Machines and Systems (ICEMS), Sapporo, Japan. 2012. pp. 1-6

[38] Nandi R, Panigrahi BK. Detection of fault in a hybrid power system using wavelet transform. In: Michael Faraday IET International Summit 2015, Kolkata, India. 2015. pp. 203-206

[39] Olivares-Galvan JC, Escarela-Perez R, González JAÁ, Jacobo JC, Guillén D, Espino-Cortés FP. Detection of interturn faults during transformer energization using wavelet transform. In: 2016 IEEE International Autumn Meeting on Power, Electronics and Computing (ROPEC), Ixtapa Mexico. 2016. pp. 1-5

[40] Gedam A, Paikrao A, Khade P. Fault diagnosis in transformer by using wavelet transform approach. In: 2018 Fourth International Conference on Advances in Electrical, Electronics, Information, Communication and Bio-Informatics (AEEICB), Chennai, India. 2018. pp. 1-4

[41] Sheshyekani K, Gharehpetian GB. A wavelet-based energy profile approach for discrimination between fault and inrush current. In: 2005 IEEE Russia Power Tech, St. Petersburg, Russia. 2005. pp. 1-5

[42] Peilin P, Guangbin D. Power quality detection and discrimination in distributed power system based on wavelet transform. In: 2008 27th

Chinese Control Conference, Kunming, China. 2008. pp. 635-638

[43] Kazemi Kargar H, Jabbari M, Zadeh SG. Inrush current identification based on wavelet transform and correlation factors. In: 2009 6th International Conference on Electrical Engineering/ Electronics, Computer, Telecommunications and Information Technology, Pattaya, Chonburi, Thailand. 2009. pp. 50-53

[44] Gafoor SA, Devi NR, Rao PVR. A transient current based bus zone protection scheme using wavelet transform. In: 2008 IEEE International Conference on Sustainable Energy Technologies, Singapore, Singapore. 2008. pp. 1195-1199

[45] Kruglova TN. Wavelet analysis for fault diagnosis of electrical machines using current signals. In: 2016 2nd International Conference on Industrial Engineering, Applications and Manufacturing (ICIEAM), Chelyabinsk, Russia. 2016. pp. 1-5

[46] Suman T, Mahela OP, Ola SR. Detection of transmission line faults in the presence of wind power generation using discrete wavelet transform. In: 2016 IEEE 7th Power India International Conference (PIICON), Bikaner, India. 2016. pp. 1-6

[47] Biswas S, Kumar K, Ghosal A, Nayak PK. Fault detection and classification for TCSC compensated transmission lines using wavelet energy. In: 2018 4th International Conference on Recent Advances in Information Technology (RAIT), Dhanbad, India. 2018. pp. 1-5

[48] Manjusree Y, Goli RK. Fault detection of a hybrid energy source integration with multiterminal transmission line using wavelet analysis. In: 2017 International Conference on Recent Trends in Electrical, Electronics and Computing Technologies (ICRTEECT), Warangal, India. 2017. pp. 120-126

[49] Wu S, Piao Z, Ji J, Li X, Hu B, Yang P. Harmonics detection in electric power systems based on wavelet packet transform. In: 2010 International Conference on Intelligent Computation Technology and Automation, Changsha, China. 2010. pp. 425-427

[50] Yang X, Xie W. Research of the power harmonic detection method based on wavelet packet transform. In: 2010 International Conference on Measuring Technology and Mechatronics Automation, Changsha City China. 2010. pp. 1112-1115

[51] Raharjo J, Panangsang O, Soeprijanto A, Zein H. Reduction in harmonic voltages on distribution system using wavelet transform. In: 2016 IEEE International Conference on Power and Energy (PECon), Melaka, Malaysia. 2016. pp. 239-243

[52] Tse NCF. Practical application of wavelet to power quality analysis. In: 2006 IEEE Power Engineering Society General Meeting, Montreal, Que., Canda. 2006. p. 5

[53] Basu M, Basu B. Analysis of power quality (PQ) signals by continuous wavelet transform. In: 2007 IEEE Power Electronics Specialists Conference, Orlando, FL, USA. 2007. pp. 2614-2618

[54] Chand P, Davari A, Liu B, Sedghisigarchi K. Feature extraction of power quality disturbances using adaptive harmonic wavelet transform. In: 2007 Thirty-Ninth Southeastern Symposium on System Theory, Macon, GA, USA. 2007. pp. 266-269

[55] Kamble S, Naik CA, Kundu P. Detection of simultaneous transient and harmonics disturbances employing discrete wavelet transform. In: 2013 Annual International Conference on Emerging Research Areas and 2013

International Conference on Microelectronics, Communications and Renewable Energy, Kanjirapally, India. 2013. pp. 1-5

[56] Tse NCF, Chan JYC, Lau W, Lai LL. Hybrid wavelet and Hilbert transform with frequency-shifting decomposition for power quality analysis. IEEE Transactions on Instrumentation and Measurement. 2012;**61**(12):3225-3233

[57] Panigrahi BK, Baijal A, Krishna Chaitanya P, Nayak PP. Power quality analysis using complex wavelet transform. In: 2010 Joint International Conference on Power Electronics, Drives and Energy Systems & 2010 Power India, New Delhi, India. 2010. pp. 1-5

[58] Praditphol N, Premrudeepreechacharn S. Classification fault events using wavelet transform and expert rules. In: 41st North American Power Symposium, Starkville, MS, USA. 2009. pp. 1-5

[59] Dasgupta A, Nath S. Detection, measurement and classification of power quality disturbances in distribution systems. In: 2009 IEEE International Advance Computing Conference, Patiala, India. 2009. pp. 917-923

[60] Yang Z, Ren G. Power quality disturbance recognition based on fitting redundant lifting wavelet packet and energy analysis. In: 2017 IEEE International Conference on Energy Internet (ICEI), Beijing, China. 2017. pp. 148-153

[61] Dekhandji FZ. Detection of power quality disturbances using discrete wavelet transform. In: 2017 5th International Conference on Electrical Engineering - Boumerdes (ICEE-B), Boumerdes, Algeria. 2017. pp. 1-5

[62] Salem ME, Mohamed A, Samad SA, Jones O. DSP implementation of power quality disturbance analysis using continuous wavelet transform. In: 2006 IEEE International Power and Energy Conference, Putra Jaya, Malaysia. 2006. pp. 94-98

[63] Eristi B, Yildirim O, Eristi H, Demir Y. A real-time power quality disturbance detection system based on the wavelet transform. In: 2016 51st International Universities Power Engineering Conference (UPEC), Coimbra, Portugal. 2016. pp. 1-5

[64] Hua L, Baoqun Z, Hong Z. Recognition and classification of power quality event in power system using wavelet transformation. In: 2008 27th Chinese Control Conference, Kunming, China. 2008. pp. 43-46

[65] Panwar A, Bisht R, Jha P. Power quality analysis using dual tree complex wavelet transform. In: 2012 Students Conference on Engineering and Systems, Allahabad, Uttar Pradesh, India. 2012. pp. 1-5

[66] Mahaddalkar SL, Shet VN. Comparative analysis of power quality using wavelets for real time implementation. In: 2016 IEEE 7th Power India International Conference (PIICON), Bikaner, India. 2016. pp. 1-4

[67] Górecki K. WPT in power quality measurement system based on digital signal processor. In: 2014 16th International Conference on Harmonics and Quality of Power (ICHQP), Bucharest, Romania. 2014. pp. 783-787

[68] Tse NCF, Zhou L. Detection of voltage variations due to distributed energy resources. In: 2009 IEEE Power & Energy Society General Meeting, Calgary, AB, USA. 2009. pp. 1-5

[69] Zheng H, Lizi Z. The factor analysis of short-term load forecast based on wavelet transform. In: Proceedings. International Conference on Power

System Technology, Kunming, China. Vol. 2. 2002. pp. 1073-1076

[70] Ji Z, Zhang P, Zhao Z. Application of wavelet neutral network and rough set theory to forecast mid-long-term electric power load. In: 2009 First International Workshop on Education Technology and Computer Science, Wuhan, Hubei, China. 2009. pp. 1104-1108

[71] Liu T. Research on the electric load forecasting and risk assessment based on wavelet neural network. In: 2009 Third International Symposium on Intelligent Information Technology Application, Shanghai, China. 2009. pp. 568-571

[72] Zhang Q, Liu T. Research on mid-long term load forecasting base on wavelet neural network. In: 2010 Second International Conference on Computer Engineering and Applications, Bali Island, Indonesia. 2010. pp. 217-220

[73] Koo B, Lee H, Park J. A study on short-term electric load forecasting using wavelet transform. In: IEEE PES Innovative Smart Grid Technologies, Europe, Istanbul, Turkey. 2014. pp. 1-6

[74] Xia C, Lei B, Rao C, He Z. Research on short-term load forecasting model based on wavelet decomposition and neural network. In: 2011 Seventh International Conference on Natural Computation, Shanghai, China. 2011. pp. 830-834

[75] Gupta S, Singh V, Mittal AP, Rani A. Weekly load prediction using wavelet neural network approach. In: 2016 Second International Conference on Computational Intelligence & Communication Technology (CICT), Ghaziabad, India. 2016. pp. 174-179

[76] Liu T. Research on the short-term electric load forecasting based on wavelet neural network. In: 2009 International Conference on Information Management, Innovation Management and Industrial Engineering, Xi'an, China. 2009. pp. 20-23

[77] Zhang Q. Research on mid-long load forecasting based on SVM and wavelet neural network. In: 2010 International Conference on Machine Vision and Human-Machine Interface, Kaifeng, China. 2010. pp. 283-287

[78] Alam MS, Chowdhury AH. Power system load switching monitoring using wavelet transform based multiresolution signal decomposition. In: International Conference on Electrical & Computer Engineering (ICECE 2010), Dhaka, Bangladesh. 2010. pp. 287-290

[79] Zhuang L, Liu H, Zhu J, Wang S, Song Y. Comparison of forecasting methods for power system short-term load forecasting based on neural networks. In: 2016 IEEE International Conference on Information and Automation (ICIA), Ningbo, China. 2016. pp. 114-119

[80] Pan J, Qi M. Study on short-term load forecasting of distributed power system based on wavelet theory. In: 2018 10th International Conference on Measuring Technology and Mechatronics Automation (ICMTMA), Changsha, China. 2018. pp. 170-173

[81] Sun W, Bai Y. Short-term load forecasting based on wavelet transform and BP neural network. In: 2011 Second International Conference on Mechanic Automation and Control Engineering, Hohhot, China. 2011. pp. 656-659

[82] Binh PTT, Hung NT, Dung PQ, Hee L. Load forecasting based on wavelet transform and fuzzy logic. In: 2012 IEEE International Conference on Power System Technology (POWERCON), Auckland, New Zeland. 2012. pp. 1-6

[83] Wen X-L, Yu C. Measurement of voltage fluctuation and flicker in electric power system based on wavelet transform. In: 2007 International Conference on Wavelet Analysis and Pattern Recognition, Beijing, China. 2007. pp. 1822-1826

[84] Pan J, Qi M. Study on short-term load forecasting of distributed power system based on wavelet theory. In: 2018 10th International Conference on Measuring Technology and Mechatronics Automation (ICMTMA), Changsha, China. 2018. pp. 170-173

Chapter 6

Wavelet Transform Applied to Internal Defect Detection by Means of Laser Ultrasound

Hossam Selim, Fernando Piñal Moctezuma, Miguel Delgado Prieto, José Francisco Trull, Luis Romeral Martínez and Crina Cojocaru

Abstract

Laser-generated ultrasound represents an interesting nondestructive testing technique that is being investigated in the last years as performative alternative to classical ultrasonic-based approaches. The greatest difficulty in analyzing the acoustic emission response is that an in-depth knowledge of how acoustic waves propagate through the tested composite is required. In this regard, different signal processing approaches are being applied in order to assess the significance of features extracted from the resulting analysis. In this study, the detection capabilities of internal defects in a metallic sample are proposed to be studied by means of the time-frequency analysis of the ultrasonic waves resulting from laser-induced thermal mechanism. In the proposed study, the use of the wavelet transform considering different wavelet variants is considered due to its multi-resolution time-frequency characteristics. Also, a significant time-frequency technique widely applied in other fields of research is applied, the synchrosqueezed transform.

Keywords: laser ultrasound, internal defect detection, wavelet transform, synchrosqueezed transform, time of flight, nondestructive testing

1. Introduction

Structural damage is a typical defect in metallic structures and components that are exposed to deformations during the manufacturing process. Such undesired physical discontinuities imply quality level affectation of the final products and even the posterior performances when subjected to complex and cyclic loadings during their service. Thus, in the last years, a more comprehensive attention has been taken to nondestructive testing (NDT) methods in order to inspect the internal characteristics of metallic components for looking for internal defects or discontinuities.

In this regard, the use of conventional Acoustic Emission (AE) transducers has the advantages of moderate cost and easy implementation, and it allows the generation of specific waveforms with a known pulse shape. Although these methods provide satisfactory results, AE transducers also show some drawbacks including the low output power, that prevents such systems from being used remotely, low frequency bandwidth range, that makes necessary the use of arrays or ultrasonic scanners increasing the system overall cost, small surface area, that prevents covering large object areas at once, and low spatial resolution in the excited volume. Ultrasonic transducers use waves with central frequencies ranging from fractions to multiples of MHz. AE analysis commonly relies on either of two schemes, pulse-echo mode or pitch-catch mode. Pulse-echo mode is more useful in applications where it is required to use only one sensor for the send/receive signals. This has some limitations on some data acquisition speed and sensor's sensitivity and size. It is also hard to recognize the location of the defects at an angle. Hence, the defect should be vertically aligned with the sensor in order to catch it. The pitch-catch mode offers more flexibility to work in both transmission and reflection modes where it is possible to more deeply investigate the ultrasonic-material interaction at different levels inside the material and extract more data concerning the defect by taking measurements at different angles. However, this technique is more expensive as it requires the use of many sensors, and the data processing is slower [1–3]. The frequency of the ultrasonic signal used affects the sensitivity and resolution of the measured defect dimensions. At higher frequencies, smaller defects can be detected more accurately. However, increasing the frequency has a negative impact on wave propagation inside the material. In other words, higher frequencies travel closer to the surface. So, the portion of the waves that penetrate to the depth of the material is reduced, thus leading to weaker possibilities to catch deeply embedded defects. Most of the available ultrasonic NDT instruments use these types of conventional AE transducers. Typically, they analyze the ultrasonic pulse's Time of Flight (TOF) through the material under test (from the transducer to the receiver) in order to identify discontinuities in the structure corresponding to potential defects.

As an alternative, photonic approaches based on laser-induced ultrasonic and optical detection showed up as valuable competitors to the conventional ultrasonic techniques in the NDT field. These techniques offer the possibility of remote transmission and detection at a much higher resolution [4, 5]. The energy carried by a laser pulse incident on an isotropic specimen is rapidly absorbed into a shallow volume of the material and creates a localized heating, which results in a thermo-elastic expansion of the material, inducing a stress wave that generates an acoustic pulse [2]. Such thermoelastic effect plays an important role in ultrasonic wave generation when the power density of the pulsed laser is lower than the ablation threshold of material. Ultrasonic waves mainly include longitudinal waves, shear waves, surface acoustic waves, and Lamb waves. Optical systems based on the laser technology can be used as well for the detection of transmitted and/or reflected acoustic waves. Several methods are implemented for this purpose. The vibration created by the acoustic wave at the surface can be optically detected using several approaches. They include optical interference techniques where a laser beam, reflected by the object surface, interferes with a reference beam. The interference fringes provide information about the crack's position and size. A Mach-Zehnder interferometer is the simplest example for how interference fringes are generated. The holographic interferometry technique is most commonly used for crack localization and flaw size determination [6]. It can detect very small details of the object under test. The optical approaches have important advantages such as the remote noncontact application, remote control, and generation of broadband frequency waves from kHz to GHz, high output power and the possibility to easily scan a larger object area at once. As an example of this performance, the work presented by Zhao et al. used this method for fatigue and subsurface crack detection [7]. Also, Erdahl discussed a valuable example of this approach to inspect multi-layered ceramic capacitors [8]. The main drawbacks of optical detection methods are their critical stability and the need for an anti-vibration setup in order to obtain reliable results, which make them very expensive and hard to apply to certain related fields.

On the other side, it is difficult to control the acoustic pulse shape as this mainly depends on the optical beam absorption properties at the material surface.

In this regard, a third approach represented by a hybrid scheme, composed by laser-ultrasonics, is considered a good trade-off in order to take advantage of both strategies, that is, the advantages of the optical system for generating artificial acoustic emission waves, as well as using a conventional ultrasonic transducer for detection. This significantly shows interesting results that overcome the drawback of the other schemes. Laser-ultrasonics offers an alternative to conventional ultrasonic techniques in the field of NDT evaluation. It allows inspection at a far distance from the object allowing the remote investigation of the test specimen without the need for a direct contact. Additionally, this technique features a broader frequency bandwidth compared with the limited bandwidth of the conventional ultrasonic transducers. Practically speaking, laser-ultrasonics covers the majority of the ultrasonic bandwidth which is important for various applications involving material characterization [2]. Indeed, the potential of this hybrid sensing scheme, combined with performative signal processing techniques, results in a promising field of study. Many researchers have made efforts to investigate the features of laser-generated acoustic waves and got substantial research achievements. For example, Zhang et al. studied empirical mode decomposition (EMD) to analyze the ultrasonic signals captured from an object that suffers from a certain defect which is followed by the Fourier transform of the selected intrinsic mode functions (IMFs) extracted from the EMD [9]. Also, Li et al. studied the laser-generated ultrasonic wave frequency characteristics in order to analyze crack effects and extract them from their generated frequency components [2]. Dixon et al. used pulsed laser-generated ultrasonics and EMAT for detecting the crack position using the B-scan study in time and frequency domains [10]. Lee discussed the ultrasonic flaw signal and technique to extract features using the fast Fourier transform and discrete wavelet transform [11]. All these studies conclude that broadband frequency components appear in the ultrasonic waves generated by the laser impulse. The Fourier Transform (FT) is the simplest and most straight forward topology for separating the frequency's components and studying their responses individually. However, it has some drawbacks since it does not allow the visualization of the temporal fingerprints of those individual frequencies. This makes it harder to figure out which frequency component corresponds to the defect. That necessitates the use of a stronger technique as the Wavelet Transform (WT) in order to analyze these frequency components and extract only those that correspond to the defect under investigation. Thus, the WT shows what frequencies are present and their impact on the time domain. Hence, it is possible to distinguish temporal and spectral behaviors, both at a time. This property helps to get more specific information about the TOF of possible reflected signals from the material with defects. Higher frequencies travel faster and closer to the surface of the object under test, compared with the lower frequencies. The wavelet technique helps to visualize the propagated frequencies at certain time instants in the ultrasonic signal life time.

Considering the nature of the acoustic emission waves generated by the laser excitation, the detection of the TOF represents a challenge that is being currently attended to by the scientific community [12]. Although some studies have exhibited the potential of the WT to analyze acoustic emission signals, the analysis and interpretation of the resulting time-frequency maps under a laser-ultrasonic scheme is still a challenge, mainly with respect to the determination of the TOF, where the error minimization is highly important. In fact, the error in determining the TOF, due to the presence of defects in the material under inspection, could become a challenge due to inconsistencies in the analysis. In this regard, the wavelet transform capabilities and some of the most recent variants exported from other fields of investigation, as the Synchrosqueezed Transform (ST), are considered. Thus, in this chapter, a defective metallic component for damage detection and visualization, through a laser-ultrasonic approach and detection of AE waves TOF, is studied. For this objective, the wavelet transform performance, as a time-frequency processing tool, and its results, are studied, compared with a promising variant called synchrosqueezed transform. This chapter is organized as follows: The theoretical basis and its suitability for the ultrasound processing of the wavelet transform and the synchrosqueezed transform are presented in Section 2. The materials and method, including the experimental setup, are explained in Section 3. The competency of the techniques and the experimental results are presented and discussed in Section 4. Finally, this chapter shows the conclusion dissemination in Section 5.

2. Theoretical considerations

2.1 Wavelet transform

The conventional Fourier series representation gives the information of frequency components in a periodic signal (inability to provide frequency information over a period of time). The simplest solution, then, is to apply the FT within a limited time interval. Thus, the time window is shifted, and frequency components are obtained using the FT. This is the principal idea of the Short Time Fourier Transform (STFT). However, due to its fixed time window, its capabilities in front of complex nonstationary signals, where frequency components vary widely over a short time interval, are reduced. The wavelet transform overcomes such limitations by introducing a scaling function, which gives a variable time window. The WT provides a variable frequency resolution unlike the FT and STFT which have a constant resolution [13].

The selection of the mother wavelet provides different characteristics of the input signal set that can emphasize certain features at the output. The flexibility of choosing the optimal mother wavelet is one of the advantages of using the WT, since the choice of the mother wavelet for a particular problem improves the signal processing capability of the technique. If the shape of the signal to be detected is known a priori, a replica of the set can be utilized as the mother wavelet function, or the mother wavelet can be chosen from a set of theoretical signals. The Mexican hat, Morlet and Daubechies4 (db4) wavelets have been proven to be efficient in improving the signal strength and reducing the noise, making the WT-based technique extremely useful for flaw detection (**Figure 1**).

The wavelet transform employs a sliding window function that is used to decompose the signal into a sum of wavelets added together. Each wavelet has finite propagation in time determined by the window size. These wavelets are limited in time, whereas sinusoidal functions, which are used for the Fourier series and Fourier transform, are continuous in the whole time range. Hence, we can use these wavelets

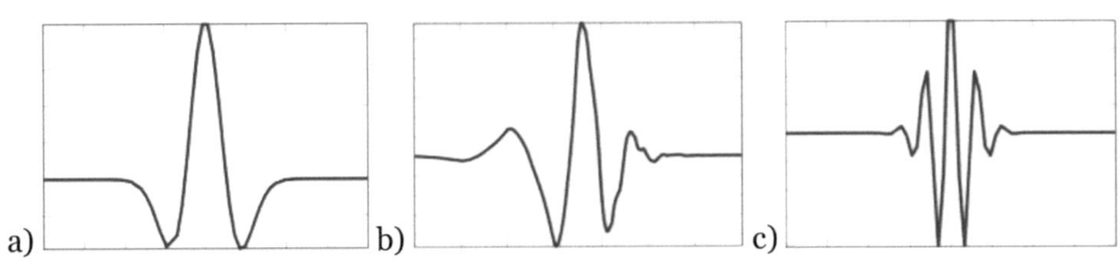

Figure 1.
Examples of wavelets used for acoustic emission processing. (a) Mexican hat. (b) Daubechies. (c) Morlet.

that can be stretched/compressed in frequency and shifted in time to correlate them with the original signal under analysis in order to determine the set of frequencies propagating at any instantaneous time to a certain level of accuracy that is still not completely accurate due to the uncertainty principle, but this accuracy is sufficient to acquire enough information about both time and frequency composition of the signal.

Assuming that a multicomponent time series signal of interest $v(t)$ can be expressed in the general form (Eq. 1):

$$v(t) = \sum_{k=1}^{K} a_k(t) \text{expcos}(2\pi i\, \phi_k(t)) \quad (1)$$

where $a_k(t)$ are the time-dependent instantaneous amplitudes, $\phi_k(t)$ denotes the instantaneous phases, and consequently, $\phi'_k(t)$ represents its instantaneous frequencies. The wavelet transform can be represented Eq. (2), where the wavelet transform Wv from the $v(t)$ signal is obtained:

$$W_v(a,b) = \frac{1}{\sqrt{|a|}} \int_{-\infty}^{\infty} v(t)\, \psi^*\left(\frac{t-b}{a}\right) dt \quad (2)$$

where ψ^* is the complex conjugation of the mother wavelet (a continuous function in both the time domain and the frequency domain). A scale factor a either stretches (a is large), or compresses (a is small) the signal, where a = ω_0/ω, ω is the angular frequency and ω_o is the angular frequency shift, while b is the signal's time shift [14–16]. The main purpose of the mother wavelet is to provide a source function to generate the daughter wavelets which are simply the translated and scaled versions of the mother wavelet.

2.2 Synchrosqueezed transform

Although, in comparison with the time-dependent Fourier transform (i.e., STFT), the achieved resolution of the time-frequency representation (TFR) by means of the wavelet analysis is certainly improved, its use still entails uncertainties on the distribution of energy for the said representation. This becomes particularly evident for nonstationary signals with a higher multimodal complexity. While it is true that these inaccuracies somehow respond to the Heisenberg-Gabor uncertainty principle [17], the fact is that they are heavily related to the choice of the wavelet function with regard to the phenomenon of the application.

In order to overcome this drawback, alternative TFR strategies have been developed. As is the case of the Wigner-Ville distribution (and their modified alternatives, e.g., Gabor-Wigner, Choi-Williams, Cohen's class, Zhao-Atlas marks, among others), despite accomplishing high resolution TFRs, their use results in additional difficulties as in the case of high computational load and artificial frequency components due to the interference between actual ones (cross-term property).

An additional TFR technique is the Hilbert-Huang Transform (HHT); by means of an Empirical Mode Decomposition (EMD) of the analyzed signal, a collection of Intrinsic Mode Functions (IMFs) is obtained, which, along with the Hilbert spectral analysis, will lead to a time-frequency depiction. Although having been successfully applied in a wide range of fields due to adaptively decomposing the signal of interest, its use also carries some drawbacks. Such is the case of a high computation load, the requirement of a stopping criterion for the EMD, the difficulty for discerning separate frequency components in narrow-band signals and a mix of modal components.

A more recent TFR framework inspired, by the adaptive approach of the HHT and the redistribution concept of the Wigner-class analysis, is the Synchrosqueezing

Transform (ST). This framework was developed with the aim to eliminate distorted interference terms while concentrating the energy on their corresponding modal components. This method, belonging to the family of the time-frequency energy reassignment, has arisen with the advantages of offering a better adaptability with regard to the signal, lesser deformation for the IF profiles, and by preserving the time, it admits an exact reconstruction formula for the constituent modal components (i.e., existence of an inverse transformation). Originally proposed for Daubechies [18] for an auditory application and revised for several authors [19–23], it works by redistributing the misallocated energy on the scale axis (due to the mother wavelet).

As aforementioned, the wavelet analysis leads to a depiction of the instantaneous frequencies $\phi'_k(t)$ of each existing component in the signal $v(t)$ by a correlation between said signal and a chosen atom (mother-wavelet), thus using a scaled and translated version of the mother-wavelet over $v(t)$. Nevertheless, under this framework is presented energy spreading over the TFR associated due to the selection of the mother-wavelet as well as for the Heisenberg-Gabor uncertainty principle, affecting the intelligibility of the analysis. The aim of the synchrosqueezed wavelet transform is to partially reassign the spread energy that occurred during the wavelet analysis for the frequency dimension only, by analyzing each component of the TFR. Therefore, it is necessary that the modal components are intrinsic mode type functions (IMT). Hence, by preserving the time dimension, it is possible to enable an inverse transformation of the obtained signal toward a time series. The wavelet synchrosqueezed transform (WST) involves the following steps. First, obtaining a wavelet transform Wv from the $v(t)$ signal following Eq. (2). Thus, ψ represents the analytic mother wavelet existing only for positive frequencies, that is, the Fourier transform of the mother-wavelet $F_{[\psi]}$ given by:

$$\hat{\psi}(\xi) = \frac{1}{2\pi\sqrt{a}}\int_{-\infty}^{\infty} \psi(t)\exp(-i\xi t)dt = 0 \quad (3)$$

for frequencies $\xi < 0$. By Plancherel's theorem, Eq. 2 can be rewritten as:

$$W_v(a,b) = \frac{1}{2\pi\sqrt{a}}\int_{-\infty}^{\infty} \hat{v}(\xi)\,\hat{\psi}^*(a\xi)\exp(ib\xi)d\xi = \frac{A}{4\pi\sqrt{a}}\hat{\psi}^*(a\,\omega_0)\exp(ib\,\omega_0) \quad (4)$$

where $\omega_0 = 2\pi f_0$ is the angular frequency of $v(t)$.

Second, extracting the IF from the wavelet transform. As each scale a of Eq. (5) corresponds to a natural frequency ξ/ω_0, satisfying the relation $a = c/\xi$ where c is the center frequency of the mother-wavelet ψ^*; it concentrates the energy of the transformation around this frequency. By supposing that the shift time b is fixed, and if $\xi = c/a$ is close, but not exactly located at the instantaneous frequency $\phi'_k(t)$, the coefficient $W_v(a,b)$ will contain some residual nonzero energy (i.e., $|Wv(a,b)|^2 > 0$), smearing the TFR. The aim of the synchrosqueezing is to remove this residual energy centered around ξ and reallocating it to a frequency location closer to its corresponding instantaneous frequency $\phi'_k(t)$. So, it is necessary to compute the instantaneous frequency of the wavelet analysis for which $W_v(a,b) \neq 0$, by the phase transformation:

$$\omega_v(a,b) = \frac{1}{i\,W_v(a,b)}\frac{\partial(W_v(a,b))}{\partial b} \quad (5)$$

Third, "squeezing" the wavelet transform over the regions where the phase transformation is constant.

During the scale-frequency mapping, that is, $(a,b) \rightarrow (\omega_v(a,b),b)$, the synchrosqueezing is applied to reassign the time-scale representation of the TF. Thus, for a fixed shift time b, the frequency reassignment $\omega_v(a,b)$ is carried out for all a scale values by means of Eq. (5), to then, for each frequency of interest ω_l, compute the synchrosqueezing by adding all values $W_v(a,b)$, where the reassigned frequency $\omega_v(a,b)$ is equal to ω_l. This is achieved by means of the mapping (for discrete values):

$$T_v(\omega_l, b) = \frac{1}{\Delta\omega} \sum_{a_k : |\omega(a_k,b) - \omega_l| \le \frac{\Delta\omega}{2}} W_v(a_k, b)\, a_k^{-3/2} (\Delta a)_k \qquad (6)$$

where $\Delta\omega = \omega_l - \omega_{(l-1)}$, $(\Delta a)_k = a_k - a_{(k-1)}$, ω_l is the lth discrete angular frequency, and a_k is the kth discrete scale point. Finally, the instantaneous angular frequency can be normalized by 2π as the IF $f = \omega / 2\pi$.

In general, the modal components from the synchrosqueezed analysis are separated well enough in the TF plane. For a given signal, if this condition is actually met, their modal components could be treated as intrinsic mode function types and their trajectories (known as wavelet ridges) can be tracked over the TF plane as their energy varies in terms of the function of time, enabling their transformation into the time domain.

3. Method and material

In order to analyze the suitability of the wavelet transform and the synchrosqueezing to extract a proper TOF related to defect location, a specific experimental bench has been arranged. The procedure is based on five steps. The first step consists of the caption of the ultrasonic signals received by the ultrasonic sensors from all considered laser scan points. These acquired signals are then processed by a noise filtering algorithm and an interpolation and bandpass filter to remove any unimportant components. The resulting A-scan signals are then ready for the next step of applying the wavelet or the synchrosqueezed transform. These transforms will generate the time frequency maps that are useful for detecting the most important propagating frequencies with respect to their times of flight. In order to further clean the signal, it is proposed to apply a signal contouring algorithm. This will help to identify the areas with uniform intensities, and the signal distribution will become clearer. It should be noted that the most important feature in this kind of algorithm is the expected time of flight for the signal. This time of flight is used later on detecting the distance between the sensor and the defect based on the speed of propagation of the ultrasonic waves (**Figure 2**).

The distance between the individual laser scan points and the receiving sensor is known a priori. In addition, the dimensions of the object under test are also known. In this regard, the time of flight of the main echoes should be equal to, or greater than, either the distance of the path from the laser direct to the sensor, or from the laser to any object boundary and reflecting back to the sensor whichever found shorter. Thus, the distance between the laser and the object boundaries is larger than the direct distance between the laser and the sensor. In addition, if there is any existing defect inside the material, this would create an internal deflection with a distance shorter than that of the object's boundaries. Hence, it is expected that the first main echo received in the analysis is due to the laser's direct propagation toward the sensor, and the second main echo, in this case, should be due to the deflections from any existing defect.

Figure 2.
The sequence flow chart of the signal processing procedure for the analysis.

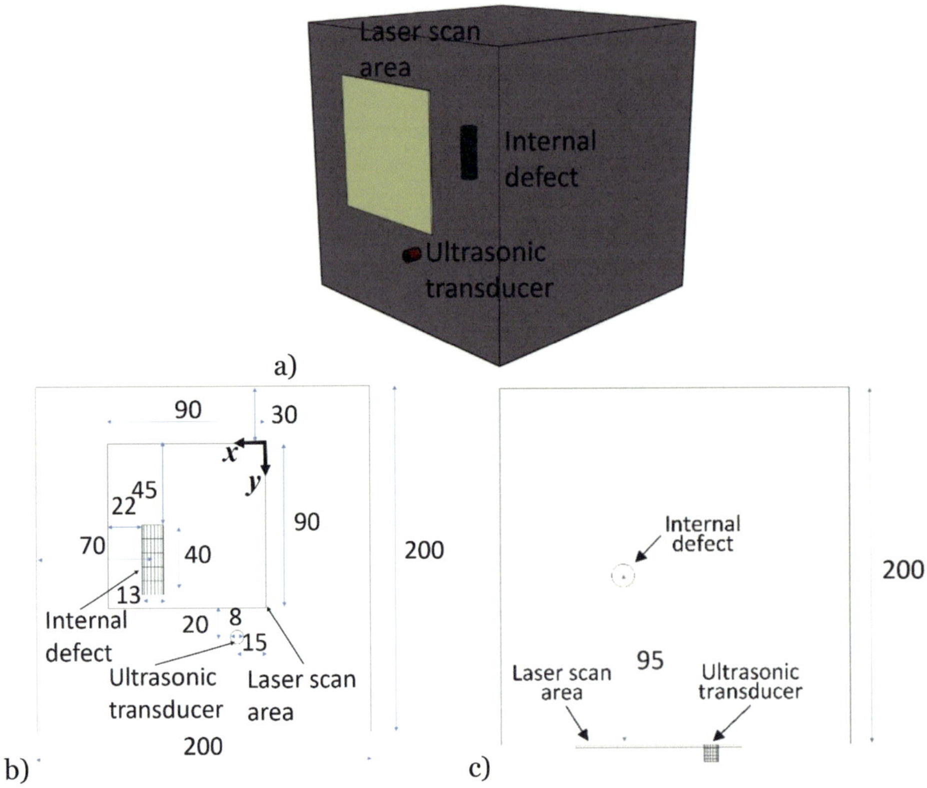

Figure 3.
Aluminum specimen, internal damage, and laser scan area. All dimensions are in millimeters. (a) Isometric view. (b) Front view. (c) Top view.

The TOF corresponding to the presence of the defect will be equal to the sum of TOF from source of excitation to the defect scatterer and the TOF from the defect scatterer to the receiving sensor. IF this TOF is converted to distance by multiplying by the longitudinal velocity of sound in the material, we can see that the position of the defect scatter would be any point at the surface of a locus ellipsoid whose two foci are the exciter and sensor positions [24].

An aluminum cube, with dimensions of 200 mm^3, and with an embedded cylindrical defect is considered to investigate the detection capabilities of the wavelet and synchrosqueezed transforms. The sample's structure and the position of the defect are shown in the next figure. The hole under investigation is the one on the top around the scan area of the laser-generated ultrasound (**Figure 3**).

4. Experimental results

This section shows the wavelet analysis of each A-scan signals detected by the sensor at positions indicated in **Table 1**. Three different types of mother wavelets are used to analyze the signal, namely the Mexican hat, Morlet, and db4. The three mother wavelets are very popular for ultrasound wave analysis due to their high correlation with the ultrasound wave form.

Figure 4 shows the cross section front view at XY plane of the cylindrical defect embedded at depth of the object. The position of the three scan points at the surface of the object are superimposed on same Figure for clarification the horizontal and

Scan point	*X* position	*Y* position	True time of flight
R1	11	51	19.65
2	31	51	18.30
3	81	51	17.10

Table 1.
Scan points considered for samples inspection.

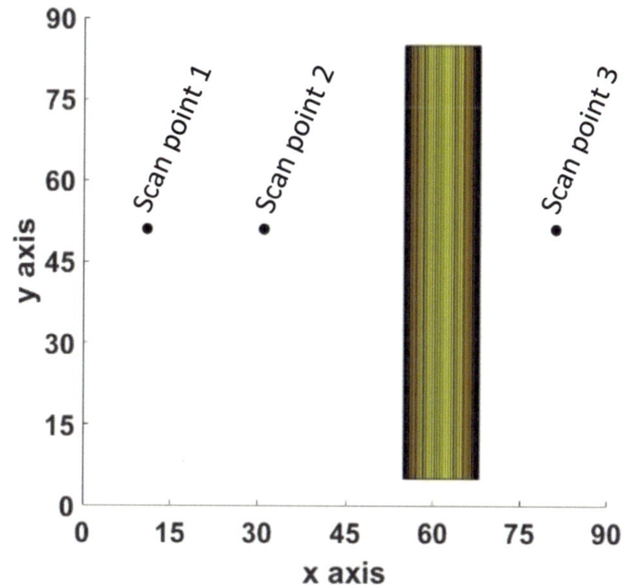

Figure 4.
Scan points considered for sample inspection and synthetic defect location.

vertical spacing from cylindrical defect. The exact data of the position of each scan point is represented in **Table 1**.

Figure 5 shows the absolute position of the defect cylinder represented by top view for the three scanning points. In addition, the locus ellipsoid estimation based on distance corresponding to TOF from exciter to defect point scatter and back to the receiving sensor for the three scanning points. It can be seen that the intersection between the ellipsoid and the cylinder happens at the point of back scattering from the defect to the sensor.

It is shown in **Figure 6** the wavelet contour map generated using Morlet WT. It is clear that the WT analysis resulted in clustering the signal into groups of segregated echoes. Each echo is governed by its intensity level, time duration, and scale levels. Scale levels are inversely proportional to the frequency spectrum. Hence, we can see at the top of the WT spectrum lies the echoes with low frequencies, while the echoes at the bottom correspond to high frequency components. Each of these echoes starts at a certain time shift, and it is clear that the start of the echo is occurring at lower frequencies with less intensity, and later the higher frequency components start to appear with their intensity level increasing. TOF of the first echo is corresponding to the direct surface propagation of the signal from the excitation point to the receiving sensor position, while TOF of the second large echo signal is corresponding to the reflected signal from the defect. It is possible to estimate the corresponding TOF based on that conclusion to be 20, 19, and 18.5 μs

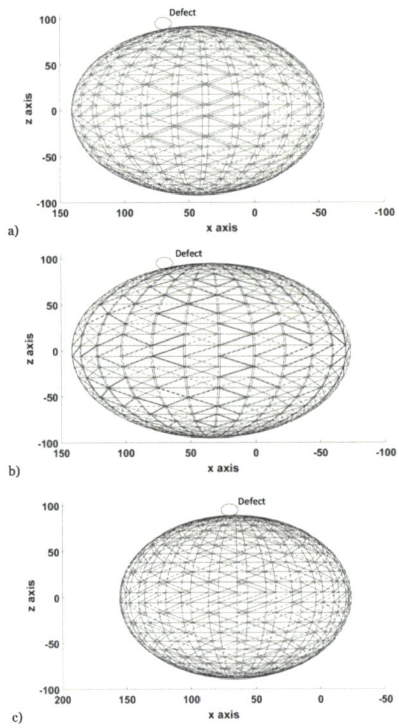

Figure 5.
Ellipsoid locus of defect position based on true time of Flight estimation resulting from the scan point and acoustic transducer positions. (a) Scan point 1. (b) Scan point 2. (c) Scan point 3.

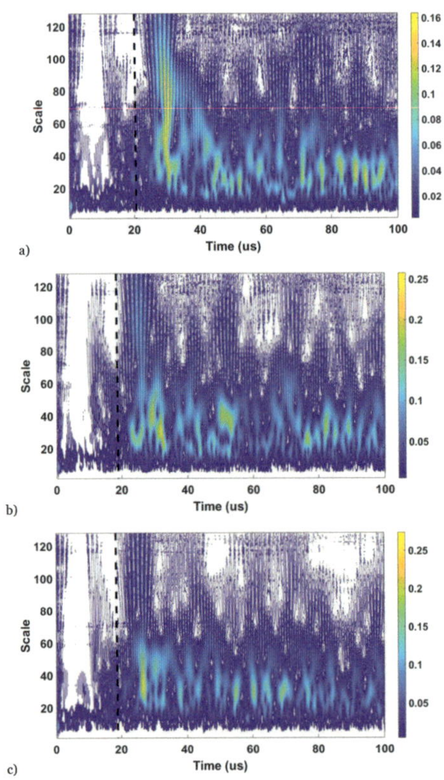

Figure 6.
Resulting wavelet contour maps with Morlet wavelet. (a) Scan point 1. (b) Scan point 2. (c) Scan point 3.

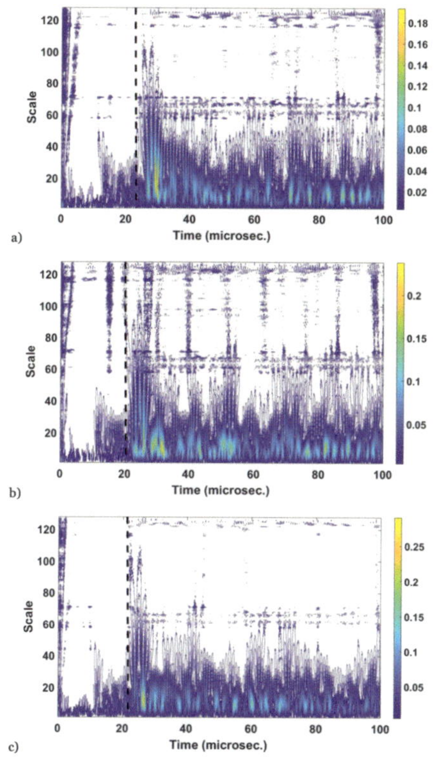

Figure 7.
Resulting wavelet contour maps with Mexican hat wavelet. (a) Scan point 1. (b) Scan point 2. (c) Scan point 3.

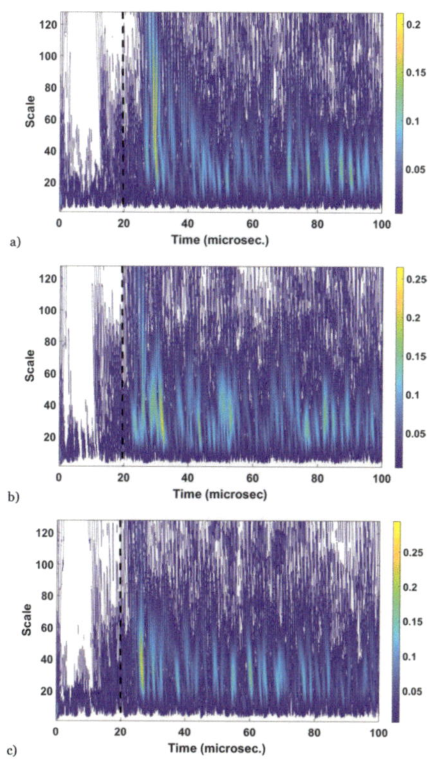

Figure 8.
Resulting wavelet contour maps with db24 wavelet. (a) Scan point 1. (b) Scan point 2. (c) Scan point 3.

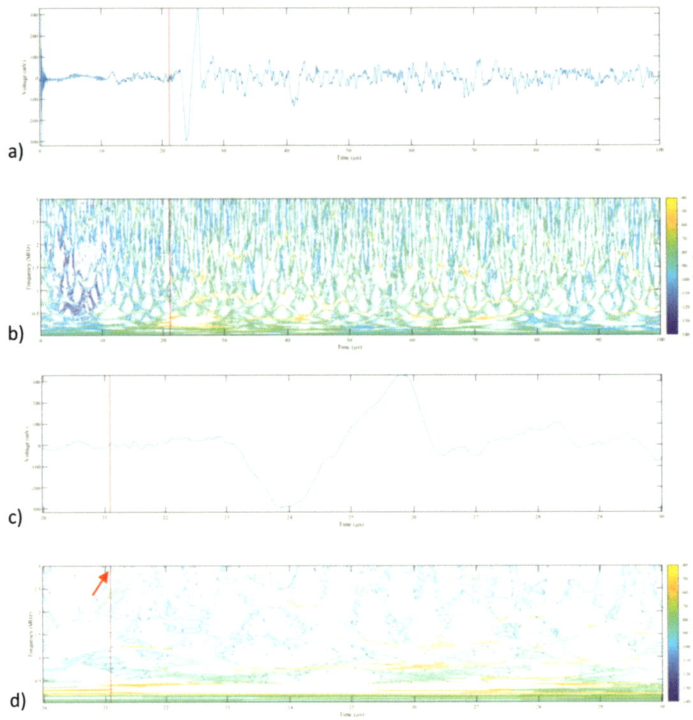

Figure 9.
Synchrosqueezing wavelet transform contour map of the scan point 1. (a) Time-based signal. (b) Synchrosqueezing transform contour map. (c) Detail of the time-based signal. (d) Detail of the synchrosqueezing transform contour map with an initial presence of acoustic activity at 21.08 μs, 2.87 MHz @−80.31 dB.

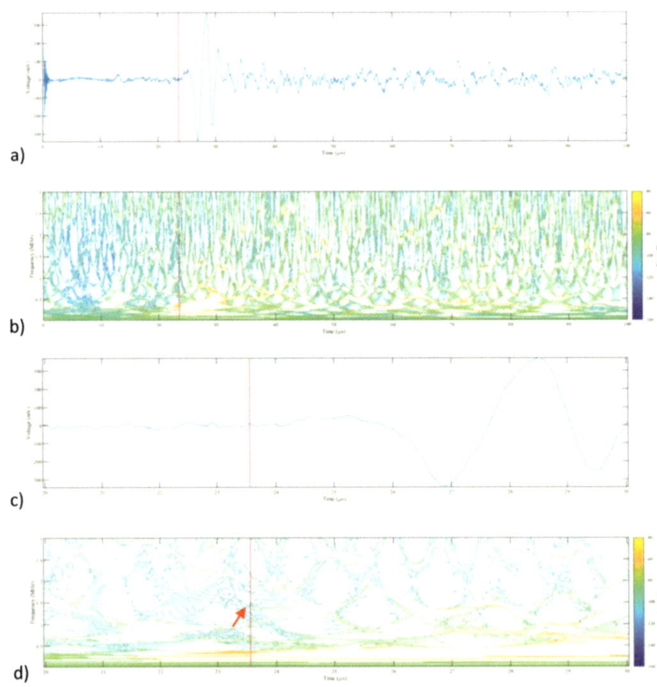

Figure 10.
Synchrosqueezing wavelet transform contour map of the scan point 2. (a) Time-based signal. (b) Synchrosqueezing transform contour map. (c) Detail of the time-based signal. (d) Detail of the synchrosqueezing transform contour map with an initial presence of acoustic activity at 23.56 μs, 2.15 MHz @−84.04 dB.

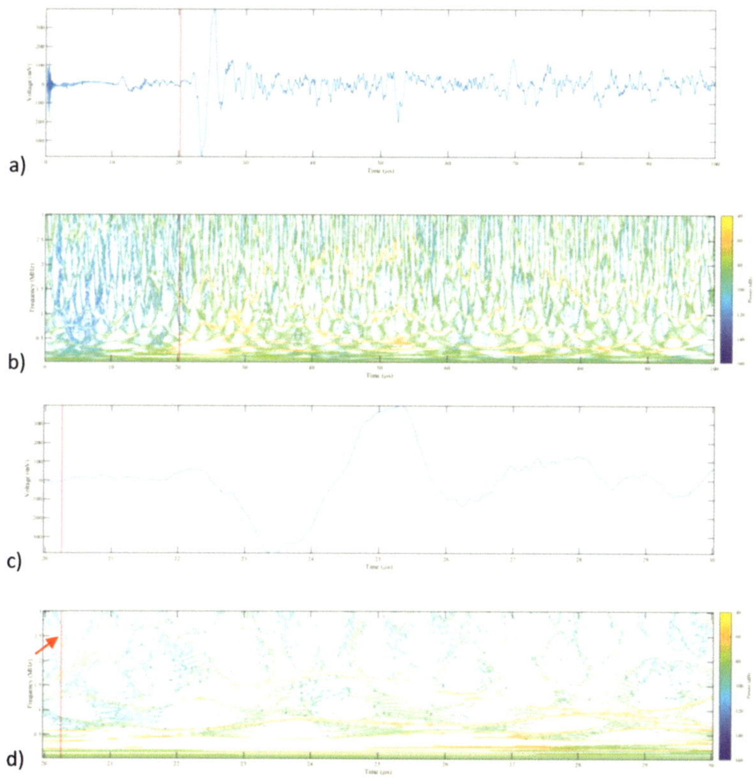

Figure 11.
Synchrosqueezing wavelet transform contour map of the scan point 3. (a) Time-based signal. (b) Synchrosqueezing transform contour map. (c) Detail of the time-based signal. (d) Detail of the synchrosqueezing transform contour map with an initial presence of acoustic activity at 20.244 μs, 2.57 MHz @−89.64 dB.

Wavelet	Scan point 1	Scan point 2	Scan point 3
Morlet	20.22	18.58	18.28
Mexican hat	23.05	19.98	20.98
Db4	21.65	18.95	18.50

Table 2.
Resulting time of flight from the internal defect in microseconds.

Wavelet	Scan point 1	Scan point 2	Scan point 3
Morlet	0.57	0.28	1.18
Mexican hat	3.40	1.68	3.88
Db4	2.00	0.65	1.40

Table 3.
Resulting time of flight error compared with the true time of flight, in microseconds.

	Scan point 1	Scan point 2	Scan point 3
Synchrosqueezing transform	21.08	23.56	20.24

Table 4.
Resulting time of flight from the internal defect in microseconds.

	Scan point 1	Scan point 2	Scan point 3
Synchrosqueezing transform	1.43	5.26	3.14

Table 5.
Resulting time of flight error compared with the true time of flight, in microseconds.

for **Figure 6a–c**, respectively. **Figure** 7 and **Figure 8** represent the same wavelet echo analysis for Mexican hat and db4 mother wavelets respectively. **Table 2** show the resulting TOF from the internal defects at three scan points with reference to the different mother wavelets while **Table 3** shows the error of the resulting TOF compared to the calculated true TOF for each of the cases in (**Table 2**). It is found that the use of the Morlet mother wavelet gives the least estimation error and it is apparently the most accurate mother wavelet to use for this kind of analysis.

For the case of the TFR that results from the application of the SSWT, it can be observed that by the accurate redistribution of the energy that compose to the signal, the obtained images achieve an improved depiction of their modal frequencies in comparison with the conventional CWT, aiding to superiorly identify the behavior of the phenomenon. Moreover, for the scope of application of this study, by identifying the first instant of time when the bi-dimensional manifold created by means of the contour mapping of the SSWT apparently becomes closed by connecting all the modal frequencies of the signal of interest, it is possible to determine the onset of said signal.

As is well known, the accurate determination of this instant of time is critical for the TOF-related methods; hence, by means of this methodology, the required precision for the onset pick is achieved when only the signal waveform is used for this purpose. Synchrosqueezing wavelet transform contour map for the scan points of interest and TOF estimation are calculated **Figures 9–11**. **Tables 4–5**. show the resulting TOF for the three scan points and Table 5 shows the corresponding error with comparison to the true TOF.

Nevertheless, considerations must be taken in order to not analyze a very small signal, this with the aim to avoid the negative effects of the Cone of Influence (COI) of the CWT, since the SSWT still leads to inaccuracies for these areas.

5. Conclusion

Indeed, the acoustic emission phenomena have been utilized as a powerful tool with the purpose to either detect, locate or assess damage for a wide range of applications. Derived from its monitoring, one of the major challenges in analyzing the resulting wavelet or synchrosqueezing transform signal is to identify and extract each generated AE event. Typically, this event detection is carried out by a thresholding approach over the raw signal. In this regard, the wavelet algorithm has resulted in a very useful and successful technique in detecting the time of flight of the acoustic emission echoes generated by defects at their corresponding frequencies. The accuracy of the algorithm was investigated experimentally using metallic structure. This algorithm is more powerful than the conventional Fourier transform algorithm. Various mother wavelets have been used to compare the correlation between the mother wavelet and the acquired A-scan signals. A mother wavelet with higher correlation would provide more accurate results. Thus, it is important to select the mother wavelet carefully to avoid misleading results. In regard with the synchrosqueezing transform, although improved resolution capabilities, the error in regard with the time of flight determination is not reduced. The Morlet wavelet is revealed as the most suitable wavelet dealing with such acoustic emission waves generated by means of LASER excitation.

Acknowledgements

This work was supported in part by the CONACyT scholarship grant number 411711, Mexico, and the Ministry of Economy and Competitiveness under the TRA2016-80472-R Research Project, Spain.

Conflict of interest

All authors declare that no conflict of interest exists at the time of manuscript submission.

Author details

Hossam Selim[1], Fernando Piñal Moctezuma[2], Miguel Delgado Prieto[2]*, José Francisco Trull[1], Luis Romeral Martínez[2] and Crina Cojocaru[1]

1 Physics Department, Technical University of Catalonia, Terrassa, Barcelona, Spain

2 Electronics Engineering Department, Technical University of Catalonia, Terrassa, Barcelona, Spain

*Address all correspondence to: miguel.delgado@mcia.upc.edu

References

[1] Park JW, Yang IY, Im KH, Hsu DK, Song SJ, Kim HJ, et al. Nondestructive evaluation of rayleigh pitch-catch contact ultrasound waves on impacted-damaged composites. Materials Science Forum. 2008;**566**:267-272. DOI: 10.4028/www.scientific.net/MSF.566.267

[2] Li J, Zhang H, Ni C, Shen Z. Analysis of laser generated ultrasonic wave frequency characteristics induced by a partially closed surface-breaking crack. Applied Optics. 2013;**52**:4179-4185. DOI: 10.1364/AO.52.004179

[3] Tiwari KA, Raisutis R, Samaitis V. Hybrid signal processing technique to improve the defect estimation in ultrasonic non-destructive testing of composite structures. Sensors. 2017;**17**(12):2858. DOI: 10.3390/s17122858

[4] Everton SK, Dickens P, Tuck C, Dutton B. Identification of sub-surface defects in parts produced by additive manufacturing, using laser generated ultrasound. In Materials Science & Technology Conference and Exhibition. 2016;**1**:141-148

[5] Shan Q, Dewhurst RJ. Surface-breaking fatigue crack detection using laser ultrasound. Applied Physics Letters. 1993;**62**:2649-2651. DOI: 10.1063/1.109274

[6] Kreis T. Application of digital holography for nondestructive testing and metrology: A review. IEEE Transactions on Industrial Informatics. 2016;**12**:240-247. DOI: 10.1109/TII.2015.2482900

[7] Zhao Y, Ma J, Liu S, Guo R, Song J, Qing Z. Laser ultrasonic technique applied to inspect fatigue crack. In: FENDT 2013–Proceedings of 2013 Far East Forum on Nondestructive Evaluation/Testing: New Technology and Application; 2013. pp. 205-8. DOI:10.1109/FENDT.2013.6635557

[8] Erdahl DS, Ume IC. Online-offline laser ultrasonic quality inspection tool for multilayer ceramic capacitors–Part I. IEEE Transactions on Advanced Packaging. 2004;**27**:647-653. DOI: 10.1109/TADVP.2004.831823

[9] Zhang Y, Yang LU, Fan J. Study on feature extraction and classification of ultrasonic flaw signals. Wseas Transactions on Mathematics. 2010;**9**:529-538

[10] Dixon S, Burrows SE, Dutton B, Fan Y. Detection of cracks in metal sheets using pulsed laser generated ultrasound and EMAT detection. Ultrasonics. 2011;**51**:7-16. DOI: 10.1016/j.ultras.2010.05.002

[11] Lee K, Estivill-Castro V. Feature extraction and gating techniques for ultrasonic shaft signal classification. Journal Applied Soft Computing. 2007. DOI: 10.1016/j.asoc.2005.05.003

[12] Zhang P, Ying CF, Shen J. Directivity patterns of laser thermoelastically generated ultrasound in metal with consideration of thermal conductivity. Ultrasonics. 1997;**35**:233-240. DOI: 10.1016/S0041-624X(96)00106-0

[13] Gómez M, Castejón C, García-Prada J. Review of recent advances in the application of the wavelet transform to diagnose cracked rotors. Algorithms. 2016;**9**:19. DOI: 10.3390/a9010019

[14] Choi J, Hong JW. Characterization of wavelet coefficients for ultrasonic signals. Journal of Applied Physics. 2010;**107**. DOI: 10.1063/1.3429087

[15] Gao RX, Yan R. Wavelets: Theory and Applications for Manufacturing. US: Springer; 2011. DOI:

10.1007/978-1-4419-1545-0. https://doi.org/10.1007/978-1-4419-1545-0

[16] Abbate A, Frankel J, Das P. Wavelet Transform Signal Processing Applied to Ultrasonics. In: Thompson D.O., Chimenti D.E. (eds) Review of Progress in Quantitative Nondestructive Evaluation. Boston, MA : Springer. 1996. https://doi.org/10.1007/978-1-4613-0383-1_97

[17] Gabor D. Theory of communication * part 1. The analysis of information. Journal of the Institution of Electrical Engineers–Part III: Radio and Communication Engineering. 1945;**93**:429-457. DOI: 10.1049/ji-3-2.1946.0074

[18] Daubechies I, Lu J, Wu H-T. Synchrosqueezed wavelet transforms: An empirical mode decomposition-like tool. Applied and Computational Harmonic Analysis. 2011;**30**:243-261. DOI: 10.1016/j.acha.2010.08.002

[19] Li C, Liang M. Time–frequency signal analysis for gearbox fault diagnosis using a generalized synchrosqueezing transform. Mechanical Systems and Signal Processing. 2012;**26**:205-217. DOI: 10.1016/j.ymssp.2011.07.001

[20] Wang Z, Ren W, Liu J. A synchrosqueezed wavelet transform enhanced by extended analytical mode decomposition method for dynamic signal reconstruction. Journal of Sound and Vibration. 2013;**332**:6016-6028. DOI: 10.1016/j.jsv.2013.04.026

[21] Thakur G, Wu H. Synchrosqueezing-based recovery of instantaneous frequency from nonuniform samples. SIAM Journal on Mathematical Analysis. 2011;**43**:2078-2095. DOI: 10.1137/100798818

[22] Jiang Q, Suter BW. Instantaneous frequency estimation based on synchrosqueezing wavelet transform. Signal Processing. 2017;**138**:167-181. DOI: 10.1016/j.sigpro.2017.03.007

[23] Thakur G, Brevdo E, Fučkar NS, Wu H-T. The Synchrosqueezing algorithm for time-varying spectral analysis: Robustness properties and new paleoclimate applications. Signal Processing. 2013;**93**:1079-1094. DOI: 10.1016/j.sigpro.2012.11.029

[24] Selim H, Delgado Prieto M, Trull J, Romeral L, Cojocaru C. Laser Ultrasound Inspection Based on Wavelet Transform and Data Clustering for Defect Estimation in Metallic Samples. Sensors. 2019;**19**:573. DOI: 10.3390/s19030573